AF888071

Henning Raach
Falling Films in Desalination

Also of Interest

Physics of Wetting.
Phenomena and Applications of Fluids on Surfaces
Bormashenko, 2017
ISBN 978-3-11-044480-3, e-ISBN 978-3-11-044481-0

Drinking Water Treatment.
An Introduction
Worch, 2019
ISBN 978-3-11-055154-9, e-ISBN 978-3-11-055155-6

Wastewater Treatment.
Application of New Functional Materials
Chen, Luo, Luo, Pang, 2018
ISBN 978-3-11-054278-3, e-ISBN 978-3-11-054438-1

Environmental Engineering.
Basic Principles
Tomašić, Zelić (Eds.), 2018
ISBN 978-3-11-046801-4, e-ISBN 978-3-11-046803-8

Henning Raach

Falling Films in Desalination

A Computational Approach

DE GRUYTER

Author
Dr. Henning Raach
Finkenweg 50
64295 Darmstadt
Germany
info@henning-raach.com

ISBN 978-3-11-059177-4
e-ISBN (PDF) 978-3-11-059233-7
e-ISBN (EPUB) 978-3-11-059185-9

Library of Congress Control Number: 2019940607

Bibliographic information published by the Deutsche Nationalbibliothek
The Deutsche Nationalbibliothek lists this publication in the Deutsche Nationalbibliografie; detailed bibliographic data are available on the Internet at http://dnb.dnb.de.

© 2019 Walter de Gruyter GmbH, Berlin/Boston
Cover image: Henning Raach
Typesetting: VTeX UAB, Lithuania
Printing and binding: CPI books GmbH, Leck

www.degruyter.com

to Ursula and Ernst Raach

Preface

The book is suited for researchers in process or chemical engineering. Its emphasis is on seawater desalination. However, falling films have a large variety of applications, e. g., in food industry, pharmaceutical industry, and in nuclear technology to mention only a few. The methods employed here are of great importance in chemical engineering, since the evaporation of a solvent is treated. Last but not least, the benefit of turbulence wires, scarcely mentioned in the literature, is examined here.

This monograph deals with evaporating saltwater falling films with and without turbulence wires treated by the means of computational fluid dynamics (CFD). This concise summary shall be explained in more detail. Here, the falling film is a thin water skin running down a vertical plane by the action of gravity. Its thickness is about 0.2 mm. The falling film is heated by the plate causing evaporation of the water at the gas-liquid interface, while the salt stays in the solution. Turbulence wires are thin wires that are immersed by the film flow. They homogenize the film preventing the formation of dry patches on the wall. Furthermore, they promote turbulence thus enhancing the heat transfer. Computational fluid dynamics relies on the power of computers. The discretized equations of the mathematical model are solved on a numerical grid, which should be fine enough in order not to neglect import features of the flow between the grid points.

The reader shall learn about the fundamental equations as well as on the computational methods to solve them. With these two goals in mind, the book is both fundamental *and* practical for a calculation engineer. She or he shall be guided through the various steps of sophistication, from the smooth film to harmonic waves up to the direct numerical simulation with the open-source CFD software OpenFOAM. (OpenFOAM is no longer restricted to the Linux operating system, but also runs with Windows 10.)

Simulating evaporating saltwater falling films is not straightforward. I do not know a software that can fulfil the task entirely. However, how shall someone reach the goal if nobody strives for it? The problem has to be attacked from various directions. With the aid of modern computers, a lot can be accomplished nowadays. Numerous ideas will be given to the reader by my own programs. She or he is also instructed how to add the energy equation to interFoam within OpenFOAM.

This is a monograph where the reader is expected to have some intermediate knowledge in fluid dynamics and programming. Some basics in C++ are of advantage, not only for OpenFOAM. However, I also give a concise survey of fluid dynamics here, which helps to gain the right level. If someone prefers to program in a language other than C++, she or he will surely find the examples easy to transform.

Some words of gratitude are appropriate here. First, I want to thank my mentor Professor Dr.-Ing. Jovan Mitrovic for all his guidance and support during my time at the Institute of Energy and Process Engineering at the University of Paderborn. Then

https://doi.org/10.1515/9783110592337-201

there are my former colleagues of whom some have become friends. The benevolent atmosphere among the partners of the European project EasyMED is still highly appreciated. Of course, I am very grateful for the support of the project by the European Union. I dedicate this book to my parents, Ursula and Ernst Raach, for their love and constant support throughout my studies and the nice time I had with them while I lived again in my hometown of Paderborn. Since 2009, I have been appreciating the good sides of Darmstadt, my adopted hometown.

Many helpers have read the manuscript and have given their comments and corrections. Thanks a lot to them! For the remaining typos and mistakes, I hold full responsibility and they will be corrected on my website:

https://henning-raach.com/book

For a motto I may cite the German translator, Dr. Erika Fuchs, who let Gyro Gearloose say, “Dem Ingeniör ist nichts zu schwör!” I translate it back:
The engineer
is without fear!

Darmstadt
April 2019

H. Raach

Contents

1 Introduction

What does this enigmatic cover display? It is a close look upon a few turbulence wires being immersed in a water falling film. As can be seen, not much happens after the first wire. However, already behind several of them a great disorder can be spotted. One topic of this book is the influence of such tripping wires, as they are also called, on the hydrodynamics of a falling film and, eventually, on the heat transfer in a desalination plant. The plate (or wall) heats the seawater skin and evaporation of the solvent takes place at the gas-liquid interface. Figure 1.1 depicts a two-dimensional model (left) and shows a small example of a real system (right). Questions like the following come up naturally: What is the optimal spacing between the wires? How far can such a system be simulated? What else can be calculated by the means of computational fluid dynamics (CFD)? What are the physical foundations? Which numerical programs may be employed?

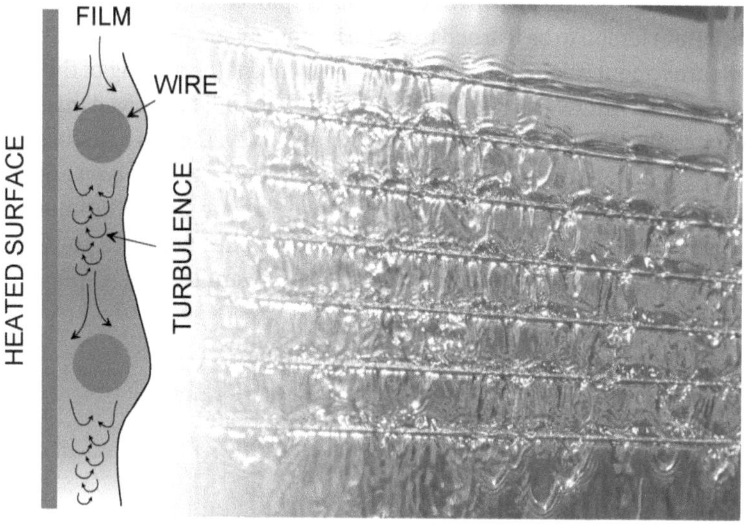

Figure 1.1: Heated falling seawater film immersing turbulence wires.

Why CFD? Numerical experiments have become the "third column of science" beside of ordinary experiments and pure theoretical reasoning. In general, CFD is faster and cheaper than ordinary experiments. Furthermore, much more data can be obtained by numerics. In contrast to theory, many rather complicated cases can be treated by it. However, although much progress has been achieved, CFD is not capable of simulating every detail of reality. Therefore, it is an art of its own to scale the problem down and to make the "right simplifications."

https://doi.org/10.1515/9783110592337-001

Three monographs on wavy falling films shall be mentioned: Alekseenko et al. [13], Chang and Demekhin [30], and Kalliadasis et al. [72]. All of these books are excellent and highly recommended for further reading. Here, the emphasis is rather on numerical calculations for chemical engineering.

After this Introduction, there is a concise chapter on seawater desalination and the European project "EasyMED," as the work for this monograph started there. Transforming saline into potable water demands a high degree of technological expertise. Chapter 2 is intended to give a survey on the most employed techniques. At the end of Chapter 2, the EasyMED desalination plant is described. Parts of this plant were to be simulated and to be optimized. Chapter 3 treats the physical foundations. There, the reader is guided through single phase fluid dynamics, two-phase flow, evaporation, and turbulence. Chapter 4 deals with the fundamentals of falling films. Chapter 5 concentrates on the numerical methods employed for this monograph. Finally, the results of the numerical experiments are shown, i. e., a description of the methodology and the presentation of pictures, tables, and diagrams. For this purpose, Chapter 6 reports on the simulations with the commercial software Star-CD. The following Chapter 7 describes how the open source software OpenFOAM was employed. The short Chapter 8 reports on an important lesson to be learned from the in-house code FS3D. In the Chapter 9, the reader learns about original programs that circumvent the solving of the full Navier–Stokes equations. Afterwards, in Chapter 10, a formula will be derived that can be helpful for a generalization of an evaporating solvent. Finally, there is a discussion and conclusion, Chapter 11. In the Appendices, technical details on *interFoam* and its extension *thinter* can be found. Furthermore, it is shown how to define turbulence wires within OpenFOAM. For this purpose, a computer aided design (CAD) software is not necessary. To the programmers among the readers, I want to be as explicit as possible. The last annex is a list of symbols.

2 Desalination

Large areas of our planet suffer from a lack and/or a poor quality of potable water. The situation will become worse due to the growing world population and global warming. On the other hand, about 70 % of the Earth is covered by the oceans containing approximately 97 % of the Earth's water. However, seawater contains salt of a concentration of 35 g/ℓ on average, whereas human beings need clean potable water of a much lower salinity (e. g., mineral water has a salinity of up to 2.5 g/ℓ). The large scale production of drinking water from seawater by the process of desalination offers a way out. This point of view is supported by the fact that about 40 % of the world's population do not live further than 100 km from the sea.

Two methods are mostly employed in industrial desalination: distillation and filtration. The separation by Reverse Osmosis (RO) is done by membranes. It is the method of choice in Europe and North America, where the seawater salinity is low and the climate is rather mild. Distillation is mostly common in the Middle East. Here, the water salinity at the shores is higher and the weather much warmer.

At the end of 2015, there were about 18,000 desalination plants worldwide, with a total installed production capacity of 86.5 million m^3/day. Approximately 45 % of this capacity is located in the Middle East and North Africa [68]. According to the International Desalination Association, desalination is practiced in 150 countries and more than 300 million people rely on desalinated water [67].

A quantitative measure of the total of organic and inorganic solutes in water is the Total Dissolved Solids (TDS). It is often expressed as parts per million (ppm) or milligrams per liter (mg/ℓ). Another measure of how many ions other than those from H_2O there are is the electrical conductivity of water. The theoretical value for pure water at 25°C is 0.055 μS/cm. However, the amount of organic solids is not accounted for by this quantity. The most common ions in typical seawater are chloride Cl^- (55 mass percent of all nonwater ions) and sodium Na^+ (30.6 %). Furthermore, there are sulfate $(SO_4)^{--}$ (7.7 %), magnesium Mg^{++} (3.7 %), calcium Ca^{++} (1.16 %), potassium K^+ (1.16 %), and carbonic acid $(CO_3)^{--}$ (0.4 %) [44].

In liquid water, the existence of a bare proton, the hydrogen nucleus H^+, is very unlikely. Rather, two water molecules separate into a hydronium H_3O^+ ion and a hydroxyl OH^- ion. In pure water at 25 °C, they have a concentration of 10^{-7} mol/ℓ. The accurate definition of the pH is minus the logarithm of the thermodynamic activity of the hydronium ion. For dilute solutions, it is a reasonable approximation to take the molar concentration as equal to the activity. Thus at 25 °C pure water has pH = 7. When acid is added, pH decreases. In the case of alkali addition, pH increases [91].

A very thorough treatment of the desalination principles can be found in Spiegler and Laird [133]. The handbook of Lorch [91] is full of instructive figures and enables the reader to get a quick access to the subject. For a more recent authoritative reference with many case studies see, e. g., El-Dessouky and Ettouney [44]. A general book on

https://doi.org/10.1515/9783110592337-002

water worldwide is the one by Hopp [66] (in German). More practical advice is given in a manual by the United States Army [140]. The American Water Work Association [16] describes the desalination technology mainly in Northern America. Voutchkov wrote two books on desalination engineering [145, 146] and Zheng on solar energy desalination technology [153]. A book completely dedicated to multistage flash desalination is by Woldai [151]. In 2006, the AAAS magazine *Science* stated that "Desalination Freshens Up" [127]. In the same magazine, an article on the future of this field was published in 2011 [45]. Of course, the Elsevier journal *Desalination* is a good source of reference and keeps up with the latest developments. Another useful journal is *Desalination and Water Treatment* by Taylor and Francis.

2.1 Reverse osmosis

The process Reverse Osmosis (RO) is illustrated in Figure 2.1. There are two containers: the right one is filled with pure water, the other with a water-salt solution. The wall between the two compartments partly consists of a semipermeable membrane which only lets the water pass through it. In the beginning, there is a flow of pure water through the membrane from the right to the left compartment. This phenomenon is called osmosis. (If there was no membrane, the salt would diffuse from the left to the right container.) As the water flows from the right to the left, the pressure rises in the left container, until there is an equilibrium and no net flow occurs. This pressure head is called the osmotic pressure. Now, we can reverse the direction of the flow of

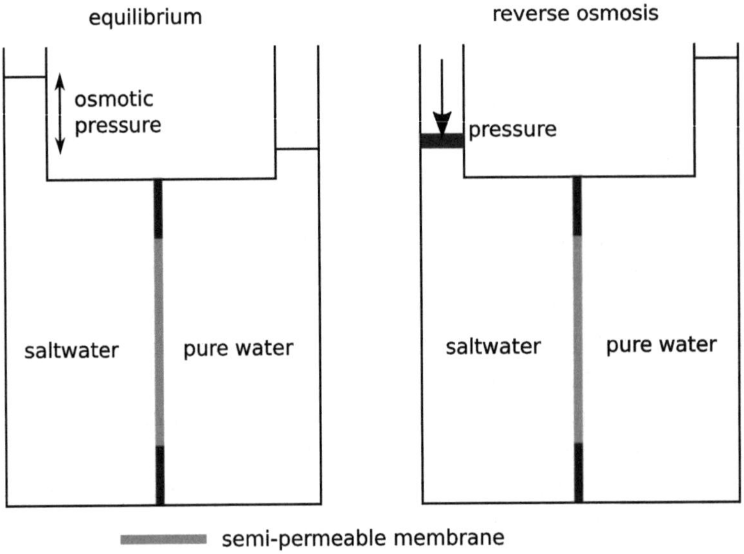

Figure 2.1: Illustration of osmotic equilibrium and reverse osmosis.

the solvent by exerting an additional pressure on the solution in the left compartment. Thus the water flows through the membrane from the left to the right. In this manner, we gain water which we shall call product. The solution on the left with the increased salinity is the so-called brine. Although the name Reverse Osmosis (RO) is well established, some experts are not happy with it. They use "hyperfiltration," as in reality the membrane lets a little amount of salt pass through. Others speak of "piezo osmosis" since the reason for the reverse flow is a pressure.

Along with the development of polymer technology came the development of suitable membranes. In the 1960s, S. Loeb and S. Sourirajan invented a salt rejecting asymmetric porous membrane made of cellulose acetate [90]. This type is called "asymmetric" on account of a very thin selective skin and an underlying porous structure giving a reasonable mechanical stability. Their membrane was the first that allowed for a high product flow rate. In 2006, the European Desalination Society established the "Sidney Loeb Award" to commemorate this important invention. Other membranes followed, e. g., those based on polyamide. The question why these materials work seems not to have been completely answered yet. It is still a formidable task to invent new membranes. On the other hand, the performance of the old materials was so satisfactory that for a long time no strong effort was made to improve the membranes of the RO plants. For an account of more recent developments see, e. g., [127] and [45].

In order to obtain a high product flow rate, the selective skin of the membrane should be as thin as possible. On one hand, it is true that the higher the pressure the higher is the flow rate through the membrane. On the other hand, higher pressure also means higher energy consumption. Furthermore, the mechanical stability of the membrane is limited resulting in a maximum pressure.

The operation of a RO plant requires a high degree of sophistication. There is the need for pre- and post-treatment of the water. Despite these measures, the membranes have to be cleaned regularly. The water to be desalinated—the feed—needs filtration to remove the larger suspended solids. Such solids that can give rise to problems can be mud and silt, organic colloids, iron corrosion products, precipitated iron, algae, bacteria, rocks, silica-like sand, precipitated manganese, and precipitated hardness. An antiscalant should be added to prevent membrane scaling, i. e., the forming of compounds like calcium carbonate, calcium sulfate, and others, which are very difficult to remove by membrane cleaning. For example, it is common use to feed acid against calcium carbonate scaling. Furthermore, biocides are added against biofouling. If chlorine is used for this purpose a dechlorination unit must be placed before certain types of membranes which are sensitive to chlorine attack. After having passed the membrane, the water usually needs some post-treatment before it can be delivered to the consumer.

Today sophisticated devices are employed, the so-called isobaric chambers. Those pressure exchangers consist of a barrel that rotates around its longitudinal axis. Longitudinal pipes are in the barrel, which are part of the high pressure and low pressure circuits. On account of the rotation, the two circuits are connected and the pressure is exchanged. Thus the feed gets a high pressure and the brine is rejected at a low pressure. Since the highest costs of a RO plant arise from exerting the pressure, a lot of energy is saved by the isobaric chambers. A more detailed account on isobaric chambers is given in the papers of MacHarg [95] and Stover [135].

It has been mentioned that RO is employed where the seawater salinity is not high and where the climate is mild. The biggest advantage of RO is the efficient use of energy. For an account of the price for RO see, e. g., Advisian [3] or a white paper by the Water Reuse Association [148]. Energy and Capital estimates a price of $0.50 per cubic meter [4]. Other techniques like distillation may be more straightforward and less sophisticated. However, these methods can only be competitive in countries where energy is cheap.

2.2 Saline distillation

Distillation is the oldest technique to produce potable water. The seawater is heated and part of the water evaporates leaving the rest in the brine. The vapor is guided into a condenser where the reverse phase transition takes place. In a simple still, energy has to be provided to pump the feed into the still, heat the seawater to the saturation temperature, provide the latent heat of evaporation, and pump the cooling water through the condenser. If no special provision is taken, the heat in the condenser is not used any further.

Thermodynamics reveals how inefficient a single-effect still is: To raise the temperature of 1 kg seawater from 20 °C to 100 °C requires an energy of 335 kJ. To evaporate the water about 2300 kJ are needed which is much more than what is needed to reach the saturation temperature. Once evaporated, the water goes to the condenser. Let us assume that the outlet temperature of the condenser is limited to 60 °C. To come up for the latent heat and the cooling down to 60 °C, $M = 14.8$ kg of cooling water at 20 °C are required, according to

$$1\,\text{kg distilled water} \cdot \left(c_p(100-60)\,\text{K} + \Delta h\right) = M\,\text{kg cooling water} \cdot c_p \cdot (60-20)\,\text{K}.$$

To make our still more efficient, part of the outlet water of the condenser shall be used as feed for the still. However, since 13.8 kg of water at 60 °C have to be discharged, only 7 % of the total energy is saved. This means that despite this provision about 93 % of the energy is “wasted”!

Special care has to be taken to make saline distillation more efficient. Two types of desalination plants are introduced in the following subsections: the multistage flash evaporation (MSF) and the multiple-effect distillation (MED). Afterwards, the European project “EasyMED” is described where a MED desalination plant was developed.

2.2.1 Multistage flash evaporation

Multistage Flash Evaporation (MSF) is the saline distillation principle used most often. A multistage flash plant is depicted in Figure 2.2. Seawater is preheated in the heat exchangers of the various stages. The seawater is further heated by condensing steam up to between 90 °C and 120 °C. The type of antiscalant dictates the upper limit for the seawater temperature. Afterwards, the feed is guided into the first stage. The pressure in this stage is below the saturation pressure with respect to the seawater temperature. From stage to stage, the pressure decreases until in the last stage about 10 kPa are reached. (The atmospheric pressure is about 100 kPa.) Thus part of the seawater evaporates in the flash chamber of each stage. The vapor is guided to the heat exchanger of the stage where it condenses. The distillate is the product. The brine is guided to the flash chamber of the next stage.

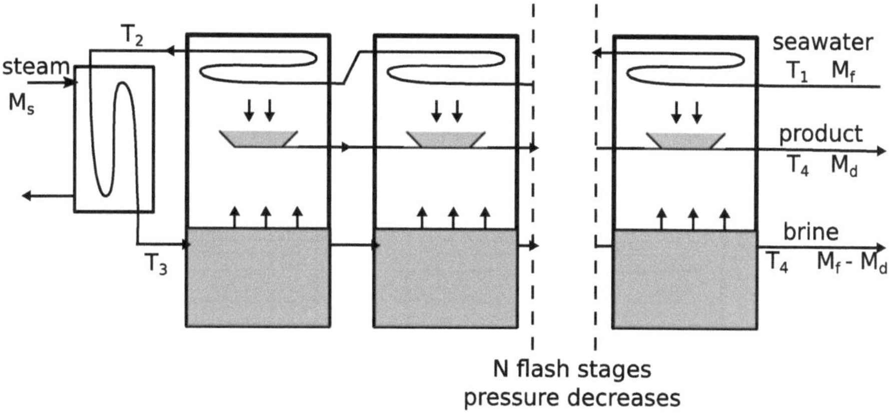

Figure 2.2: Simplified flow diagram of a multistage flash evaporator.

An important measure for the efficiency of a MSF plant is the Gained Output Ratio (GOR) also referred to as Performance Ratio (PR): It yields the mass of the distillate, M_d, divided by the mass of the heating steam, M_s. If T_1 is the initial seawater temperature, T_2 the seawater temperature after the preheating, and T_3 the maximal seawater temperature, the following two equations hold:

$$M_s \Delta h = M_f c_p (T_3 - T_2), \tag{2.1}$$

and

$$M_d \Delta h = M_f c_p (T_2 - T_1), \tag{2.2}$$

where $\Delta h = h_V - h_L$ denotes the latent heat, M_f the mass of the feed, and c_p the specific heat of the seawater. The boiling point elevation due to the increasing salt concentration has been neglected. From equations (2.1) and (2.2), the GOR is easily obtained:

$$\mathrm{GOR} = \frac{M_d}{M_s} \approx \frac{T_2 - T_1}{T_3 - T_2}. \tag{2.3}$$

In equation (2.3), it is assumed that the latent heat of the steam and the vapor are approximately the same.

2.2.2 Multi-effect distillation

A simplified model of a multi-effect distillation (MED) plant is displayed in Figure 2.3. The main idea behind it is that external thermal energy has to be supplied only to the first effect. From there, the vapor is guided to the next effect where it condenses thus releasing the latent heat to produce new vapor. This vapor is lead to the next effect where it condenses and so on. The temperature and pressure decrease from effect to effect. Furthermore, as follows from Figure 2.3, the brine from the previous effect is the feed to the next effect thus causing a boiling point elevation due to the rising salt concentration. This provision avoids a new preheating of the feed.

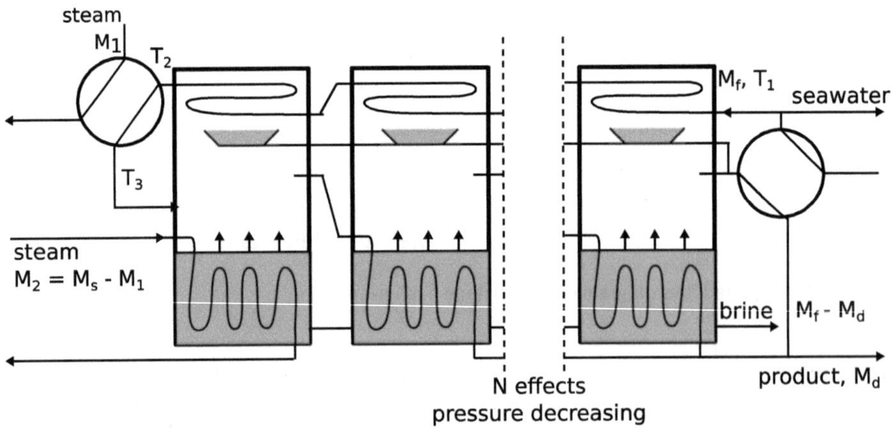

Figure 2.3: Simplified flow diagram of a multi-effect distillation plant.

Similar to a MSF plant, the seawater feed is preheated in the condenser, starting in the last effect with the lowest temperature and continuing until the feed leaves the first effect. Let T_1 denote the seawater temperature before the preheating and T_2 that after the preheating. Assume that a mass M_1 of external steam is needed to raise the feed temperature to the evaporation temperature T_3 of the first effect. Then the following equation holds:

$$M_1 \Delta h = M_f c_p (T_3 - T_2), \tag{2.4}$$

where Δh is the latent heat, M_f the mass of the feed, and c_p the specific heat of the seawater. If M_s is the total steam mass and M_2 the mass of the steam available for boiling in the hottest effect, then

$$M_2\Delta h = (M_s - M_1)\Delta h = M_s\Delta h - M_f c_p(T_3 - T_2). \tag{2.5}$$

In equation (2.5), M_1 from equation (2.4) has been taken.

Imagine an ideal MED plant where no energy is lost between the effects. Hence M_2 of distillate is produced in each of the N effects. We obtain for the total distillate mass, M_d,

$$M_d\Delta h = NM_2\Delta h = N[M_s\Delta h - M_f c_p(T_3 - T_2)]. \tag{2.6}$$

As before, the latent heat of the steam and the vapor are assumed to be approximately the same. From equation (2.6), we obtain the Gain Output Ratio, GOR,

$$\mathrm{GOR} = \frac{M_d}{M_s} = N - \frac{N(T_3 - T_2)M_f c_p}{M_s\Delta h}. \tag{2.7}$$

Since the latent heat Δh is a very large quantity, it can be seen from equation (2.7) that the GOR of an *ideal* MED plant is only a bit less than the number of effects.

$$\boxed{\mathrm{GOR} \approx N} \tag{2.8}$$

Supplied with this basic knowledge on multi-effect distillation we can more easily understand the EasyMED project.

2.2.3 The EasyMED project

In the European project EasyMED, partners from France, Italy, and Germany developed a new MED desalination plant which is easy to operate and which makes use of energy efficiently. The main elements are metallic plates: On one side, a seawater falling film runs down which is heated by the latent heat of the condensing vapor on the other side of the plate. Figure 2.4 shows cross-sectional views. The vapor leaves the evaporator through holes in the frame to a condenser of the next effect. It is guided by a pressure gradient from one effect to the next effect. Since the pressure in the cells is below the atmosphere pressure, spacers are inserted between the plates. Grids are attached to the spacers. For the falling films in the evaporators, these grids fulfil two purposes: On the one hand, they homogenize the film preventing it from building dry patches on the heated plate. On the other hand, the grid wires promote turbulence thus enhancing the heat transfer also at lower Reynolds numbers. This is the reason why these wires are often called "turbulence wires." They are also known as "tripping wires" or "tripwires".

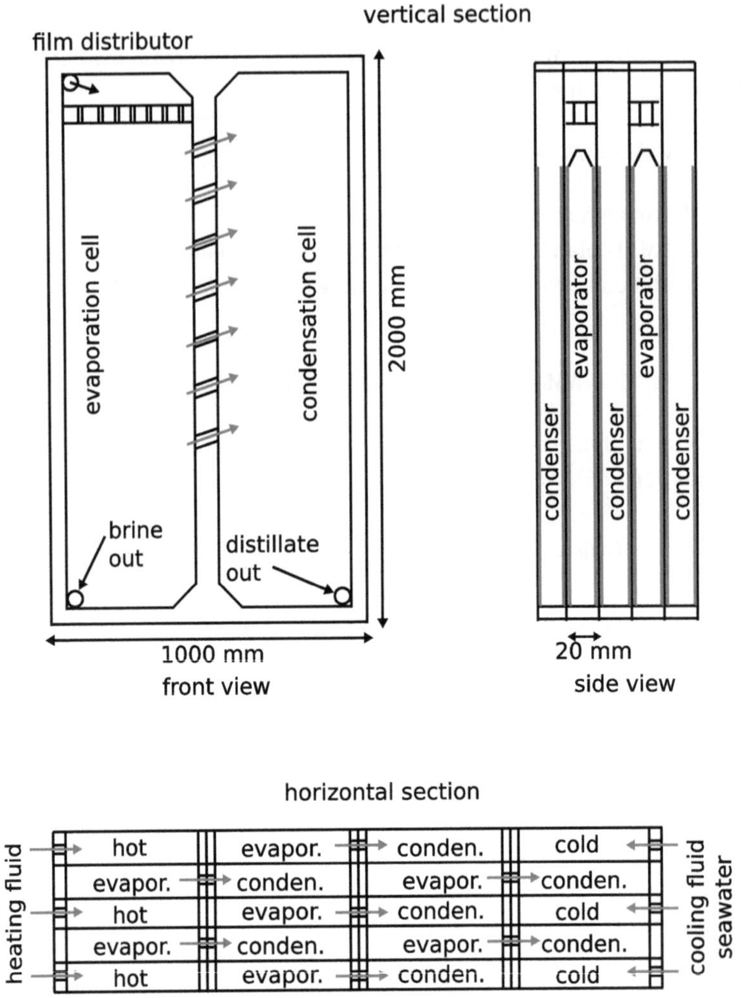

Figure 2.4: Cross-sectional views of the EasyMED desalination plant.

A substantial improvement potential for the desalination process was expected for the falling film evaporators. A couple of questions arose: How can the heat and mass transfer in these cells be modeled and simulated? What is the optimum geometry of the grid in order to achieve the maximum evaporation rate? The answer to these questions is the subject of this monograph.

Let us summarize some of the experimental facts: The first step was the testing of a one-effect plate evaporator under laboratory conditions [70]. It produced between $0.3\,m^3/d$ and $0.4\,m^3/d$ of distilled water. The distillate conductivity was below $20\,\mu S/cm$. This fact confirmed that no salted water droplets were carried over by the vapor stream. The evaporation rate ranged between 10 % and 50 %. The mean heat transfer coefficient between the heating cell and the evaporating film was about 1100 $Wm^{-2} K^{-1}$.

The next step was to build a three-effect laboratory unit and to test it with synthetic saltwater (tap water added with NaCl) [121]. The average production of distilled water was 0.95 m^3/d. The conductivity was even below 12 $\mu S/cm$. A thermal efficiency of 77 % was achieved together with a GOR $\approx$ 2.6. The temperature in the hottest effect was about 65 °C; the temperature drop from one effect to the next was about 5 °C.

Finally, the practical feasibility of such a multi-effect plate distillation plant was demonstrated with real seawater in the harbor of La Spezia, Italy [71]. A production rate of 3 m^3/d was obtained. Again the quality of the distillate was very good. The evaporation rate ranged between 8 % and 13 %. The gain output ratio was about GOR $\approx$ 2.3. The fact that the GOR was not close to 3 was attributed to the addition of cold seawater to the brine of the previous effect. Thus the inlet temperature was about 10 °C below the boiling point. It is expected that in an industrial unit with 10 effects and about 50 cells in parallel a much higher GOR and a production of 200 m^3/d can be achieved.

Last but not least, the corrosion resistance tests by our Italian partner shall be mentioned here [59]. Their material studies were also part of the EasyMED project. Among other investigations, they tested stainless steel and titanium. Both materials showed satisfactory results provided that an antiscalant and a biocide had been mixed to the seawater.

3 Physical foundations

3.1 Single phase flow

In this section, the fundamental equations of single phase fluid dynamics are rather concisely repeated. For an exposition on an elementary level see, e. g., Munson et al. [103]. For a thorough treatment at the level of this monograph, Anderson [17] is recommended.

3.1.1 Four different descriptions of fluid flow

Dealing with fluid dynamics, everyone will presumably have noticed that the basic equations appear in different forms. The reason for this is that fluid flow can be looked upon in different ways: One can observe the flow through a fixed volume or the observer can move with the flow. The size of the control volume is also important: It is either finite or infinitesimal. These possibilities all have a strong influence on the form of the equation. A finite control volume V leads to an integral equation, whereas an infinitesimal volume dV requires a differential equation. Furthermore, the observer at rest is guided to an equation in conservation form. For a volume that moves with the flow and that always consists of the same fluid particles, an equation in nonconservation form is more adequate. Although their forms are different, mathematically all of these four equations are equivalent. Thus the engineer is free to choose the most convenient form. The fixed volume approach is often referred to as Euler frame, whereas the moving observer is called Lagrange frame. In one phase, fluid dynamics for the Euler picture seems to be more often used. For a treatise that is solely dedicated to Lagrangian fluid dynamics see, e. g., the monograph by Bennett [22].

3.1.2 The substantial derivative

It is essential to know the substantial derivative to read the equations deduced from the Lagrangian point of view. It is a differential operator written as D/Dt. It can be interpreted as *the temporal change of a quantity in an infinitesimal fluid element moving with the flow.*

In order to see how the substantial derivative can be calculated, let us consider the temporal derivative of the temperature T, which we assume to be a function of space and time. According to the chain rule,

$$\frac{dT}{dt} = \frac{\partial T}{\partial t} + \frac{\partial T}{\partial x}\frac{dx}{dt} + \frac{\partial T}{\partial y}\frac{dy}{dt} + \frac{\partial T}{\partial z}\frac{dz}{dt}. \tag{3.1}$$

https://doi.org/10.1515/9783110592337-003

The term dx/dt is the x component of the observer's velocity. Similarly, dy/dt is a velocity in the y direction and dz/dt is a velocity in the z direction. Let us choose as an "observer" a fluid element moving with the flow. Its velocity in Cartesian coordinates shall be written as

$$\boldsymbol{U} = u\boldsymbol{i} + v\boldsymbol{j} + w\boldsymbol{k}, \tag{3.2}$$

where $\boldsymbol{i}, \boldsymbol{j}, \boldsymbol{k}$ are the unit vectors in the x, y, and z direction, respectively. Thus the temporal derivative becomes the substantial derivative:

$$\frac{DT}{Dt} = \frac{\partial T}{\partial t} + u\frac{\partial T}{\partial x} + v\frac{\partial T}{\partial y} + w\frac{\partial T}{\partial z}. \tag{3.3}$$

Equation (3.3) is a special, coordinate dependent form of the substantial derivative. In order to get a form independent of the coordinate system, let us introduce the Nabla operator ∇, which is a differential vector operator. In Cartesian coordinates, it is

$$\nabla = \boldsymbol{i}\frac{\partial}{\partial x} + \boldsymbol{j}\frac{\partial}{\partial y} + \boldsymbol{k}\frac{\partial}{\partial z}. \tag{3.4}$$

Thus the substantial derivative of the temperature can be rewritten as

$$\frac{DT}{Dt} = \frac{\partial T}{\partial t} + (\boldsymbol{U} \cdot \nabla)T. \tag{3.5}$$

It is instructive to understand each term on the right-hand side of equation (3.5). The partial temporal derivative yields the change of the temperature that happens *locally* at a fixed point. The second term is the change of the temperature due to *convection*, i. e., the flow through a certain point. Let us consider an example from every day life: It is summer and the weather is hot. You go into a tree's shade to get cooler. This temperature change is caused by a movement and would be described by the second "convective" term. You decide to stay in the shadow and sleep for a while. During your sleep, the shadow moves until you do not lie in a shadow any more. This temperature change is due to a local heating and would be expressed by the partial temporal derivative, the first term.

The substantial derivative of the temperature is an example. Of course, the differential operator can also be applied to quantities like the density ρ or the velocity $\boldsymbol{U}$.

3.1.3 The divergence of the velocity

Let δV be a very small volume moving with the flow. Then the divergence of the velocity can be expressed as

$$\nabla \boldsymbol{U} = \frac{1}{\delta V}\frac{D(\delta V)}{Dt}. \tag{3.6}$$

It can be interpreted as *the temporal change of a small volume moving with the flow per unit volume*. For a derivation see, e. g., Anderson [17].

We are already in the position to understand the meaning of $\nabla \boldsymbol{U} \equiv 0$, which is to appear later. The volume does not change with time along the flow. Since from a nonrelativistic point of view the mass is conserved, the density ρ must be a constant. Einstein's theory of relativity is not taken into account in most engineering applications, for it especially applies to velocities near the speed of light (special relativity) and at very large masses (like the Sun's) that perceivably bend space-time (general relativity).

3.1.4 The continuity equation

The continuity equation is a balance equation that expresses the conservation of mass. The continuity equation is going to be displayed in four different forms according to the four different ways to look upon a flow.

From a fixed infinitesimal control volume, one is led to the differential equation in conservation form:

$$\frac{\partial \rho}{\partial t} + \nabla \cdot (\rho \boldsymbol{U}) = 0. \tag{3.7}$$

Either what flows in, flows out, or the density inside the control volume changes.

From a fluid element moving with the flow, the differential equation in non-conservation form is directly obtained:

$$\frac{D\rho}{Dt} + \rho \nabla \cdot \boldsymbol{U} = 0. \tag{3.8}$$

In equation (3.8), the substantial derivative gives us a hint that we are dealing with a moving reference frame.

From a fixed control volume of finite size one is directly guided to the integral equation in conservation form:

$$\frac{\partial}{\partial t} \int_V \rho dV + \int_A \rho \boldsymbol{U} \cdot \boldsymbol{dA} = 0. \tag{3.9}$$

In equation (3.9), A denotes the control surface which is the boundary surface of the control volume V.

If the finite control volume is moving with the flow, always consisting of the same particles, the continuity equation is written as an integral equation in non-conservation form:

$$\frac{D}{Dt} \int_V \rho dV = 0. \tag{3.10}$$

Again, the substantial derivative hints at a movement.

All of these equations can be transformed into each other. Thus their mathematical equivalence is proven [17].

Finally, let us consider the important case of an incompressible fluid. (A liquid is approximately incompressible or a gas at low Mach number.) This means that the density ρ does not change. From equation (3.7), it is easily deduced that

$$\boxed{\nabla \boldsymbol{U} \equiv 0}\,. \tag{3.11}$$

The divergence of the velocity has to vanish everywhere if the density is constant. Equation (3.11) is known as the "incompressible continuity equation."

3.1.5 The Navier–Stokes equations

The Navier–Stokes equations are the momentum equations. The physical principle behind them is Newton's axiom $\boldsymbol{F} = m \cdot \boldsymbol{a}$ or, more precisely,

$$\frac{d\boldsymbol{P}}{dt} = \sum_i \boldsymbol{F}_i, \tag{3.12}$$

where $\boldsymbol{P} = m \cdot \boldsymbol{U}$ is the momentum of a fluid element with mass m. Since the mass of a fluid element is hard to capture, we are going to deal with the momentum per unit volume $\rho \cdot \boldsymbol{U}$.

The stress tensor τ_{ij} needs to be introduced, before the momentum equations are written. It yields the force per unit area in the direction j acting on the area perpendicular to the i axis; see Figure 3.1 for a shear stress and a normal stress.

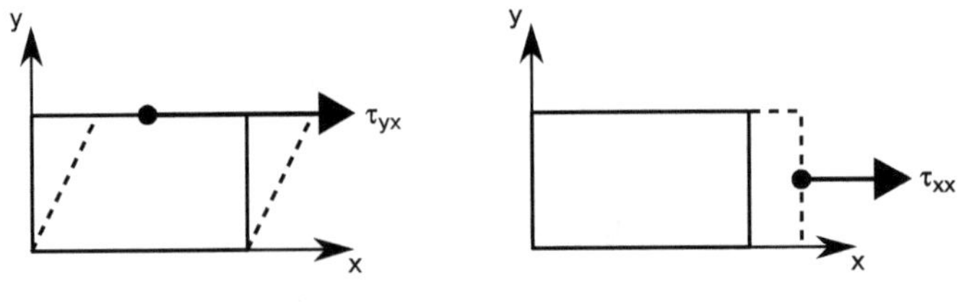

Figure 3.1: A shear stress and a normal stress.

Let us consider an infinitesimal fluid element moving with the flow. As it always consists of the same particles, its mass m is constant and can be taken outside the derivative of the left-hand side of equation (3.12) which becomes

$$m\frac{D\boldsymbol{U}}{Dt} = \sum_i \boldsymbol{F}_i. \tag{3.13}$$

If both sides of equation (3.13) are divided by the instantaneous volume δV of the fluid element, one is directly led to the differential momentum equations in non-conservation form:

$$\rho\frac{Du}{Dt} = -\frac{\partial p}{\partial x} + \frac{\partial \tau_{xx}}{\partial x} + \frac{\partial \tau_{yx}}{\partial y} + \frac{\partial \tau_{zx}}{\partial z} + \rho f_x \tag{3.14}$$

$$\rho\frac{Dv}{Dt} = -\frac{\partial p}{\partial y} + \frac{\partial \tau_{xy}}{\partial x} + \frac{\partial \tau_{yy}}{\partial y} + \frac{\partial \tau_{zy}}{\partial z} + \rho f_y \tag{3.15}$$

$$\rho\frac{Dw}{Dt} = -\frac{\partial p}{\partial z} + \frac{\partial \tau_{xz}}{\partial x} + \frac{\partial \tau_{yz}}{\partial y} + \frac{\partial \tau_{zz}}{\partial z} + \rho f_z \tag{3.16}$$

Let us have a closer look at equation (3.14), the x momentum equation. On the left-hand side, there is the density multiplied by the acceleration in the x direction. On the right-hand side, there is the sum of the forces acting in the x direction. First, there are surface forces, i. e., a reverse pressure gradient and the normal stress and shear stresses on account of the viscosity. (The stress tensor shall be substituted a bit later.) Then there is the body force $\boldsymbol{f}$ per unit mass which in most applications is just the gravitational acceleration $\boldsymbol{g}$. Equations (3.15) and (3.16) are completely analogous for the y and z direction, respectively.

The differential momentum equations in conservation form are obtained with the aid of the continuity equation:

$$\frac{\partial(\rho u)}{\partial t} + \nabla\cdot(\rho u\boldsymbol{U}) = -\frac{\partial p}{\partial x} + \frac{\partial \tau_{xx}}{\partial x} + \frac{\partial \tau_{yx}}{\partial y} + \frac{\partial \tau_{zx}}{\partial z} + \rho f_x \tag{3.17}$$

$$\frac{\partial(\rho v)}{\partial t} + \nabla\cdot(\rho v\boldsymbol{U}) = -\frac{\partial p}{\partial y} + \frac{\partial \tau_{xy}}{\partial x} + \frac{\partial \tau_{yy}}{\partial y} + \frac{\partial \tau_{zy}}{\partial z} + \rho f_y \tag{3.18}$$

$$\frac{\partial(\rho w)}{\partial t} + \nabla\cdot(\rho w\boldsymbol{U}) = -\frac{\partial p}{\partial z} + \frac{\partial \tau_{xz}}{\partial x} + \frac{\partial \tau_{yz}}{\partial y} + \frac{\partial \tau_{zz}}{\partial z} + \rho f_z \tag{3.19}$$

In the remainder of this book, we shall assume that the fluid is Newtonian, i. e., the rate of strain is assumed to be proportional to the shear stress. This is a good approximation for fluids dealt with in this monograph: water, vapor, and air. Thus we can specify the stress tensor:

$$\tau_{xx} = \hat{\mu}(\nabla\cdot\boldsymbol{U}) + 2\mu\frac{\partial u}{\partial x} \tag{3.20a}$$

$$\tau_{yy} = \hat{\mu}(\nabla\cdot\boldsymbol{U}) + 2\mu\frac{\partial v}{\partial y} \tag{3.20b}$$

$$\tau_{zz} = \hat{\mu}(\nabla\cdot\boldsymbol{U}) + 2\mu\frac{\partial w}{\partial z} \tag{3.20c}$$

$$\tau_{xy} = \tau_{yx} = \mu\left[\frac{\partial v}{\partial x} + \frac{\partial u}{\partial y}\right] \tag{3.20d}$$

$$\tau_{xz} = \tau_{zx} = \mu\left[\frac{\partial w}{\partial x} + \frac{\partial u}{\partial z}\right] \tag{3.20e}$$

$$\tau_{yz} = \tau_{zy} = \mu\left[\frac{\partial w}{\partial y} + \frac{\partial v}{\partial z}\right] \tag{3.20f}$$

The coefficient μ is called the dynamic viscosity. For the other coefficient, $\hat{\mu}$, Stokes assumed

$$\hat{\mu} = -\frac{2}{3}\mu, \tag{3.21}$$

which is often employed, but not strictly proved.

By means of the continuity equation for an incompressible fluid, the Navier–Stokes' equations now become

$$\frac{\partial(\rho u)}{\partial t} + \nabla \cdot (\rho u \boldsymbol{U}) = -\frac{\partial p}{\partial x} + \mu \nabla^2 u + \rho f_x, \tag{3.22}$$

$$\frac{\partial(\rho v)}{\partial t} + \nabla \cdot (\rho v \boldsymbol{U}) = -\frac{\partial p}{\partial y} + \mu \nabla^2 v + \rho f_y, \quad \text{and} \tag{3.23}$$

$$\frac{\partial(\rho w)}{\partial t} + \nabla \cdot (\rho w \boldsymbol{U}) = -\frac{\partial p}{\partial z} + \mu \nabla^2 w + \rho f_z. \tag{3.24}$$

Here, we assume the dynamic viscosity μ to be constant.

3.1.6 The energy equation

The energy equation expresses that energy is conserved. Consider an infinitesimal fluid element moving with the flow. Then we can describe the energy equation by the following balance:

$$\begin{matrix} \text{temporal change} & & \text{heat flux} & & \text{rate of work} \\ \text{of energy inside} & = & \text{to the interior} & + & \text{by volume and} \\ \text{fluid element} & & \text{of the element} & & \text{surface forces} \end{matrix} \tag{3.25}$$

This leads to the differential energy equation in nonconservation form:

$$\begin{aligned} \rho \frac{D}{Dt}\left(e + \frac{U^2}{2}\right) = \rho Q &+ \frac{\partial}{\partial x}\left(\lambda \frac{\partial T}{\partial x}\right) + \frac{\partial}{\partial y}\left(\lambda \frac{\partial T}{\partial y}\right) + \frac{\partial}{\partial z}\left(\lambda \frac{\partial T}{\partial z}\right) \\ &- \frac{\partial(up)}{\partial x} - \frac{\partial(vp)}{\partial y} - \frac{\partial(wp)}{\partial z} \\ &+ \frac{\partial(u\tau_{xx})}{\partial x} + \frac{\partial(u\tau_{yx})}{\partial y} + \frac{\partial(u\tau_{zx})}{\partial z} \\ &+ \frac{\partial(v\tau_{xy})}{\partial x} + \frac{\partial(v\tau_{yy})}{\partial y} + \frac{\partial(v\tau_{zy})}{\partial z} \\ &+ \frac{\partial(w\tau_{xz})}{\partial x} + \frac{\partial(w\tau_{yz})}{\partial y} + \frac{\partial(w\tau_{zz})}{\partial z} + \rho \boldsymbol{f} \cdot \boldsymbol{U} \end{aligned} \tag{3.26}$$

In equation (3.26), e denotes the specific inner energy, Q the specific volumetric heat supply (e. g., by radiation), λ the heat conductivity, p the pressure, and $\boldsymbol{f}$ the specific volume force. (We do not need to specify the stress tensor, for the terms containing its components shall be neglected soon.) Notice that equation (3.26) is the conservation equation for the total energy, i. e., the sum of inner and kinetic energy. The equation for the inner energy e reads

$$\begin{aligned}\rho\frac{De}{Dt} = \rho Q &+ \frac{\partial}{\partial x}\left(\lambda\frac{\partial T}{\partial x}\right) + \frac{\partial}{\partial y}\left(\lambda\frac{\partial T}{\partial y}\right) + \frac{\partial}{\partial z}\left(\lambda\frac{\partial T}{\partial z}\right)\\ &- p\left(\frac{\partial u}{\partial x} + \frac{\partial v}{\partial y} + \frac{\partial w}{\partial z}\right) + \tau_{xx}\frac{\partial u}{\partial x} + \tau_{yx}\frac{\partial u}{\partial y} + \tau_{zx}\frac{\partial u}{\partial z}\\ &+ \tau_{xy}\frac{\partial v}{\partial x} + \tau_{yy}\frac{\partial v}{\partial y} + \tau_{zy}\frac{\partial v}{\partial z} + \tau_{xz}\frac{\partial w}{\partial x}\\ &+ \tau_{yz}\frac{\partial w}{\partial y} + \tau_{zz}\frac{\partial w}{\partial z}\end{aligned} \tag{3.27}$$

The energy equation in conservation form is obtained with the aid of the continuity equation:

$$\begin{aligned}\frac{\partial(\rho e)}{\partial t} + \nabla\cdot(\rho e\boldsymbol{U}) = \rho Q &+ \frac{\partial}{\partial x}\left(\lambda\frac{\partial T}{\partial x}\right) + \frac{\partial}{\partial y}\left(\lambda\frac{\partial T}{\partial y}\right) + \frac{\partial}{\partial z}\left(\lambda\frac{\partial T}{\partial z}\right)\\ &- p\left(\frac{\partial u}{\partial x} + \frac{\partial v}{\partial y} + \frac{\partial w}{\partial z}\right) + \tau_{xx}\frac{\partial u}{\partial x} + \tau_{yx}\frac{\partial u}{\partial y} + \tau_{zx}\frac{\partial u}{\partial z}\\ &+ \tau_{xy}\frac{\partial v}{\partial x} + \tau_{yy}\frac{\partial v}{\partial y} + \tau_{zy}\frac{\partial v}{\partial z} + \tau_{xz}\frac{\partial w}{\partial x} + \tau_{yz}\frac{\partial w}{\partial y} + \tau_{zz}\frac{\partial w}{\partial z}\end{aligned} \tag{3.28}$$

A comparison between equation (3.27) and equation (3.28) reveals that the right-hand side has remained unchanged. Equation (3.28) shall be subject to some simplifications: The fluid is assumed to be incompressible. Furthermore, energy changes due to radiation and viscosity shall be neglected. The specific heat c_V is supposed to be constant so that

$$e = c_V T. \tag{3.29}$$

Thus from equation (3.28), we obtain an equation for the temperature:

$$\boxed{\frac{\partial(\rho c_V T)}{\partial t} + \nabla\cdot(\rho c_V T\boldsymbol{U}) = \lambda\nabla^2 T} \tag{3.30}$$

In equation (3.30), the heat conductivity λ is assumed to be constant.

3.1.7 Thermal diffusion

Diffusion is the transport by the motion of the molecules. However, we are not interested to track the positions and velocities of every molecule at a microscopic level.

We rather would like to know the result on macroscopic quantities like the temperature T. For this purpose, we assume the fluid to be at rest, macroscopically. For $\boldsymbol{U} = 0$, equation (3.30) becomes

$$\frac{\partial(\rho c_V T)}{\partial t} = \lambda \nabla^2 T. \tag{3.31}$$

As we did with λ, let us assume that the density and the specific heat are constant. We define the thermal diffusivity D_{th} as the quotient

$$D_{\text{th}} = \frac{\lambda}{\rho c_V}. \tag{3.32}$$

Then we obtain the thermal diffusion equation

$$\boxed{\frac{\partial T}{\partial t} = D_{\text{th}} \nabla^2 T}\,. \tag{3.33}$$

The diffusivity has the unit m^2/s. Thus D_{th} and the kinematic viscosity ν have the same unit. (The similarities of momentum, heat, and mass transport are nicely brought together in the textbook by Bird, Stewart, and Lightfoot [24].) The quotient of these two diffusivities is called Prandtl number:

$$\text{Pr} = \frac{\nu}{D_{\text{th}}} \tag{3.34}$$

Thermal diffusion is also known as *heat conduction*. The heat flux vector depends on the temperature gradient:

$$\boxed{\boldsymbol{q} = -\lambda \nabla T} \quad \text{Fourier's law of heat conduction} \tag{3.35}$$

Notice the minus sign. The heat flows from regions with higher temperature to cooler volumes.

In most textbooks, the line of argumentation is usually the other way around. First, Fourier's law is introduced, then the diffusion equation, and finally the energy equation. Again, Bird, Stewart, and Lightfoot [24] are highly recommended for a thorough treatment of transport phenomena. There, the reader also learns more on the connections between microscopic and macroscopic point of views. The interrelations between statistical and thermal physics are well explained by Reif [120].

3.1.8 Mass transport

In this monograph, the seawater is treated as a *binary mixture*. One component is called "salt" and the other is "water." The mass fraction of the salt is the *salinity* defined by

$$S = \frac{\rho_{\text{salt}}}{\rho_{\text{salt}} + \rho_{\text{water}}}. \tag{3.36}$$

As a good approximation, we can neglect ρ_{salt} in the denominator:

$$S \approx \frac{\rho_{\text{salt}}}{\rho_{\text{water}}}, \quad \text{since} \quad \rho_{\text{salt}} \ll \rho_{\text{water}}. \tag{3.37}$$

A common definition of S is "kg(salt)/kg(water)." The salinity is dimensionless. More often it is given as "g(salt)/kg(water)." However, in Chapter 2 the salinity has the unit g/ℓ, which is approximately the same, since 1 kg of water has a volume of about 1 liter.

The transport equation for S is very similar to the one for the temperature, equation (3.30):

$$\frac{\partial S}{\partial t} + \nabla \cdot (S\boldsymbol{U}) = D_s \nabla^2 S \tag{3.38}$$

D_s is the diffusivity of the salt in the water. If the water is at rest, equation (3.38) becomes a diffusion equation:

$$\frac{\partial S}{\partial t} = D_s \nabla^2 S \tag{3.39}$$

The kinematic viscosity divided by D_s is called the *Schmidt number*:

$$\text{Sc} = \frac{\nu}{D_s} \tag{3.40}$$

In order to complete the set of numbers built from our three different kinds of diffusivities:

$$\text{Lewis number} \quad \text{Le} = \frac{D_{\text{th}}}{D_s} \tag{3.41}$$

Analogous to Fourier's law of heat conduction, there is Fick's law. It says that the diffusive mass flux has the opposite direction of the salinity gradient. This makes sense. The diffusion of the salt is from a volume with high salinity to a region with smaller S.

3.2 Two phase flow

In this section, the reader is introduced to essential results on two phase flow that are needed to understand the state of the art in CFD. For further studies on this subject, we refer to Aris [18], Gatignol and Prud'homme [56], Slattery et al. [130], Kolev [78, 79], and Edwards et al. [43]. We are first concerned with definitions since this subject requires a high degree of mathematical sophistication. We mostly rely on the notation of Edwards et al. [43] whose book is very pedagogical and highly recommended to the novice in this subject. However, at the end of this section considerable simplifications are made for CFD.

The most common and most powerful model of an interface between two immiscible fluids is a two-dimensional (2D) surface embedded in 3D space. In the mathematical language of differential geometry, such a surface is called a “manifold.” The change between the two fluids is not so abrupt in reality. There is rather an interfacial layer of a certain extent in a third dimension. Edwards et al. [43] show in their monograph how a 2D model can be justified from the consideration of an interfacial layer. So we shall proceed with the 2D surface model.

With the identity matrix $\boldsymbol{I}$ and the surface normal vector $\boldsymbol{n}$, we can build the dyadic surface idemfactor:

$$\boldsymbol{I}_s = \boldsymbol{I} - \boldsymbol{nn} \tag{3.42}$$

The dyadic product $\boldsymbol{nn}$ is a matrix, calculated according to

$$(\boldsymbol{nn})_{ij} = n_i \cdot n_j \quad (i,j = 1,2,3). \tag{3.43}$$

In the literature, also the notation $\boldsymbol{n} \otimes \boldsymbol{n}$ can be found.

The analogue to the Nabla operator, ∇, in 3D space, is the surface gradient operator, ∇_s,

$$\nabla_s = \boldsymbol{I}_s \cdot \nabla \tag{3.44}$$

When we are dealing with interfaces, in most cases we are concerned with curved surfaces. The surface curvature dyadic is defined as

$$\boldsymbol{b} \equiv -\nabla_s \boldsymbol{n} \tag{3.45}$$

With the aid of the mutually perpendicular unit vectors in the respective directions of the principal axes of curvature, $\boldsymbol{e}_1$ and $\boldsymbol{e}_2$, an alternative to equation (3.45) can be

$$\boldsymbol{b} = \kappa_1 \boldsymbol{e}_1 \boldsymbol{e}_1 + \kappa_2 \boldsymbol{e}_2 \boldsymbol{e}_2 \tag{3.46}$$

In equation (3.46), κ_1 and κ_2 are called the principal curvatures of the surface. The mean curvature, H, is defined by

$$H \overset{\text{def}}{=} -\frac{1}{2}\nabla_s \cdot \boldsymbol{n} = \frac{1}{2}(\kappa_1 + \kappa_2) \tag{3.47}$$

The Gaussian or total curvature, K, can be calculated by

$$K \equiv \kappa_1 \kappa_2 \tag{3.48}$$

To learn more on curvature, see a book on elementary differential geometry, e. g., Pressley [114]. (Differential geometry in three-dimensional space is elementary to a mathematician.)

The velocity of a point at the interface shall be denoted by $\boldsymbol{U}_S$. In the case of a 2D interface, a so-called material surface, there is no ambiguity in such a definition.

In the literature, a distinction is drawn between two types of interfacial viscosities. The surface shear viscosity, μ^S, is a measure for the resistance against tangential stretching. The interfacial dilatational viscosity, κ^S, refers to a deformation of the interface in the normal direction. The surface viscosities are zero in the case of a pure substance without the presence of surfactants [130]. The reader interested in rheological techniques to measure these two quantities can consult, e. g., Edwards et al. [43].

Still one definition is necessary. Let $\boldsymbol{a}$, $\boldsymbol{b}$, $\boldsymbol{c}$, and $\boldsymbol{d}$ be vectors. The double-dot product of two dyads $\boldsymbol{ab}$ and $\boldsymbol{cd}$ yields the following scalar:

$$\boldsymbol{ab} : \boldsymbol{cd} = (\boldsymbol{b} \cdot \boldsymbol{c})(\boldsymbol{a} \cdot \boldsymbol{d}) \tag{3.49}$$

Now, finally, we are in the position to understand the equations for the surface forces which are usually distinguished by the directions of their actions. There is the normal surface force, $\boldsymbol{S}_n$, and the tangential one, $\boldsymbol{S}_S$. We are going to write down the expressions for so-called Newtonian surfaces. The name arises due to a similarity of the surface stress tensor and the stress tensor of a Newtonian fluid [43]. For a detailed account of non-Newtonian surfaces, cf. Slattery et al. [130]. So, the normal component of the surface force, $\boldsymbol{S}_n$, can be written as

$$\begin{aligned} \boldsymbol{S}_n = &- \boldsymbol{F}^S \cdot \boldsymbol{nn} - 2H\sigma\boldsymbol{n} \\ &- 2\mu^S\boldsymbol{n}(\boldsymbol{b} - 2H\boldsymbol{I}_S) : \nabla_S\boldsymbol{U}_S - 2H\boldsymbol{n}(\kappa^S + \mu^S)\nabla_S \cdot \boldsymbol{U}_S \end{aligned} \tag{3.50}$$

The expression for the tangential component, $\boldsymbol{S}_S$, reads

$$\begin{aligned} \boldsymbol{S}_S = &- \boldsymbol{F}^S \cdot \boldsymbol{I}_S - \nabla_S\sigma - (\kappa^S + \mu^S)\nabla_S\nabla_S \cdot \boldsymbol{U}_S \\ &- \mu^S\{\boldsymbol{n} \times \nabla_S[(\nabla_S \times \boldsymbol{U}_S) \cdot \boldsymbol{n}] + 2(\boldsymbol{b} - 2H\boldsymbol{I}_S) \cdot (\nabla_S\boldsymbol{U}_S) \cdot \boldsymbol{n}\} \end{aligned} \tag{3.51}$$

In equations (3.50) and (3.51), $\boldsymbol{F}^S$ denotes the net external force. From equation (3.50), the term $-2H\sigma\boldsymbol{n}$ shall be called "the Laplace force." It describes an effect of a constant surface tension σ when the mean curvature H of the interface is unequal to zero. The Laplace surface force has a smoothing effect on a curved surface. Let us keep it in mind:

$$\boxed{\boldsymbol{F}_{\text{Laplace}} = -2H\sigma\boldsymbol{n}} \tag{3.52}$$

Its modeling will be the topic of Section 5.5.

From equation (3.51), the expression $-\nabla_S\sigma$ shall be called "the Marangoni force." It arises when there are differences in the surface tension. This can happen due to different temperatures at the interface or in the presence of varying surfactant concentration. The Marangoni surface force induces tangential currents that tend to balance the differences. Since it will appear again in Chapter 4, we write it down separately:

$$\boldsymbol{F}_{\text{Marangoni}} = -\nabla_S\sigma \tag{3.53}$$

The other terms in the equations (3.50) and (3.51) have a factor μ^s or κ^s. Thus they originate from the surface viscosities. In computational fluid dynamics, CFD, only the Laplace force is usually taken into account. The Marangoni force is subject to CFD research; see, e. g., the Ph. D. thesis of Alke [15]. The terms due to the surface viscosities are neglected.

In this section, no phase change occurred at the interface. The phenomenon of evaporation shall be treated next.

3.3 Evaporation

A brief account of the thermodynamical meaning shall be given first. Then the boundary conditions that have to be applied to the interface shall be stated.

Evaporation is the phase transition from the liquid to the gaseous state. Depending on a control parameter phase transitions are classified as discontinuous or continuous [21, 109]. In the case of fluid dynamics, this control parameter usually is the density. Thus below the critical point (for water: $T_c = 647.4\,\text{K}$, $p_c = 22.1\,\text{MPa}$) evaporation is a discontinuous phase transition, whereas near the critical point it can be regarded as continuous. A phase diagram for water is depicted in Figure 3.2. There, the vapor pressure curve is highlighted starting at the triple point and ending at the critical point. This curve has to be crossed if evaporation is to occur. For two phases of a pure substance being in equilibrium, there is one degree of freedom according to Gibb's phase rule. Normally, this is chosen to be the temperature ("saturation temperature") or the pressure ("saturation pressure"). In the case of a discontinuous phase

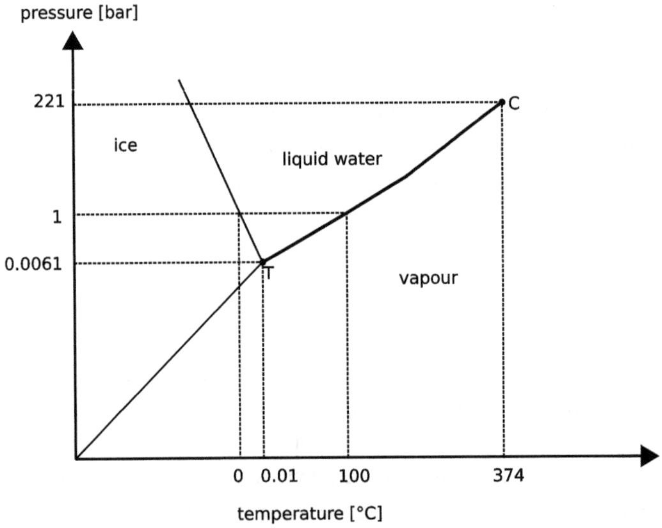

Figure 3.2: Phase diagram of water (sketch).

transition, the enthalpy of the vapor is much higher than the enthalpy of the liquid. This difference, called "latent heat," has to be supplied.

This book deals with saltwater. Therefore, it shall be mentioned here that the higher the salt concentration, the higher is the evaporation temperature ("boiling point elevation").

For a more recent treatment of evaporation, experimental and theoretical see, e. g., Badam [20]. In the remainder of this section, the question shall be pursued how evaporation can be modeled in fluid dynamics and what boundary conditions are to be applied to the gas-liquid interface. Gatignol and Prud'homme [56] distinguish many cases, among other assumptions that of evaporation near and far from equilibrium. Slattery et al. [130] state the general equations for fluid dynamics with an interface and give jump conditions for mass, momentum, energy, and entropy. The model for two phase flow with evaporation outlined here is that of Krahl et al. [80], which is the result of an interdisciplinary collaboration with the aim of a CFD simulation.

First, the conservation of mass requires symmetric relations for the mass flux, $\dot{m}$, at the liquid and gas side of the interface (subscripts "l" and "g"):

$$\dot{m} = \rho_l(\boldsymbol{U}_l \cdot \boldsymbol{n} - U_s) \tag{3.54}$$

$$\dot{m} = \rho_g(\boldsymbol{U}_g \cdot \boldsymbol{n} - U_s) \tag{3.55}$$

Furthermore, it is assumed that the tangential component of the velocity is continuous on the phase boundary ("No slip"):

$$\|\boldsymbol{U}\| = \boldsymbol{U}_l - \boldsymbol{U}_g = (\boldsymbol{U}_l \cdot \boldsymbol{n} - \boldsymbol{U}_g \cdot \boldsymbol{n})\boldsymbol{n} \qquad \text{on the interface} \tag{3.56}$$

The expression $\|\Psi\|$ should be read as "jump of the quantity Ψ on the interface." From equations (3.54), (3.55), and (3.56), a jump condition for the velocity follows:

$$\|\boldsymbol{U}\| = \left(\frac{1}{\rho_l} - \frac{1}{\rho_g}\right)\dot{m}\boldsymbol{n} \tag{3.57}$$

The momentum jump condition is

$$\begin{aligned} 2H\sigma\boldsymbol{n} &= \|-p\boldsymbol{n} + \boldsymbol{\tau} \cdot \boldsymbol{n} - \rho\boldsymbol{U}(\boldsymbol{U} \cdot \boldsymbol{n} - U_s)\| \\ &= \|-p\boldsymbol{n} + \boldsymbol{\tau} \cdot \boldsymbol{n}\| - \dot{m}\|\boldsymbol{U}\| \\ &= \|-p\boldsymbol{n} + \boldsymbol{\tau} \cdot \boldsymbol{n}\| + \left(\frac{1}{\rho_g} - \frac{1}{\rho_l}\right)\dot{m}^2\boldsymbol{n} \end{aligned} \tag{3.58}$$

On the left-hand side of equation (3.58), only the Laplace force is taken into account.

Krahl et al. assume a continuous temperature on the phase boundary:

$$\|T\| = 0 \tag{3.59}$$

The mass flux depends on the heat flux, $\boldsymbol{q}$, and on the enthalpy difference, $\Delta h = h_g - h_l$:

$$\|\boldsymbol{q}\| \cdot \boldsymbol{n} = \|-\lambda \partial_{\boldsymbol{n}} T\| = \Delta h \cdot \dot{m} \quad \Rightarrow \quad \dot{m} = \|-\lambda \partial_{\boldsymbol{n}} T\| / \Delta h \tag{3.60}$$

In equation (3.60), the symbol $\partial_{\boldsymbol{n}}$ denotes the normal derivative. The partial pressure of the vapor, p_v, has to be taken into account if there is an inert component in the gaseous phase. For an ideal gas, it can be determined by

$$p_v = \frac{\rho_v}{M} k_B T \tag{3.61}$$

In equation (3.61), M is the mass of a vapor molecule and $k_B = 1.381 \cdot 10^{-23}$ J/K is the Boltzmann constant. Concerning the temperature on the free surface, T_s, Krahl et al. calculate it from a fixed-point equation:

$$T_s = T_{\mathrm{sat}}(p_v(\rho_v, T_s)), \tag{3.62}$$

where T_{sat} is the saturation temperature. There are some comments on this procedure [80]. First, if the fluids are assumed to be incompressible, the partial pressure must not be taken from the pressure in the Navier–Stokes equations. Thus p is inconsistent in a thermodynamic meaning. Second, a nonvanishing evaporation rate requires some distance from equilibrium. Thus we are dealing with an evaporation near equilibrium.

In order to close the model of Krahl et al., a transport equation and a boundary condition for the vapor density are needed. The transport equation is just an advection-diffusion equation:

$$\frac{\partial \rho_v}{\partial t} + \boldsymbol{U} \cdot \nabla \rho_v - D_v \nabla^2 \rho_v = 0, \tag{3.63}$$

where D_v is the diffusivity. The boundary condition for the interface reads

$$D_v \partial_{\boldsymbol{n}} \rho_v = \left(\frac{\rho_v}{\rho_g} - 1 \right) \dot{m} \tag{3.64}$$

Note that there is no diffusion at the phase boundary if the mass flux is zero or the partial density/pressure of the inert component vanishes.

In a desalination plant, the presence of air is essential for the evaporation. If $p_v = p_g$, the liquid would boil! What is the difference between evaporation and boiling? Evaporation is slower and only occurs at the gas-liquid interface.

3.4 Turbulence

Although most flows in nature and engineering are turbulent, the phenomenon of turbulence is regarded as the number one unsolved problem of classical physics. There is no common definition that every expert would agree upon since it is still the subject of active research. However, we shall not be intimidated by the subject but rather briefly summarize what is known, how to model and simulate turbulence, and even propose a new model for free surfaces.

There is vast literature on turbulence. We shall concentrate here on more recent monographs. The mathematically inclined reader will find the book by Pope [110] very useful. A good physical approach for scientists and engineers is given in Davidson [35]. For the novice, the informal introduction by Tsinober [137] is recommended. A monograph by Bernard and Wallace [23] treats analysis, measurement, and prediction. Three books, the topic of which is mainly modeling and simulation, are those by Durbin [42], Gatski [57], and Hewitt [64]. Finally, lecture notes on transition and turbulence control can be found in Gad-el-Hak [52].

3.4.1 What is turbulence?

The phenomenon shall be approached by a simple example with a basic explanation. A sophisticated definition shall be given at the end of this section.

Consider a horizontal circular pipe through which a fluid is flowing. The exact solution by Hagen and Poiseuille with a parabolic velocity profile exists for laminar, unidirectional flow. Reynolds investigated the deviations from the Hagen–Poiseuille solution. He found out that, if the average velocity, $\overline{\boldsymbol{U}}$, is high, the tube diameter, d, large, and the kinematic viscosity $\nu = \mu/\rho$ small, then the flow tends to be turbulent. Reynolds combined the three factors to form a number which is named after him:

$$\mathrm{Re} = \frac{\overline{\boldsymbol{U}} \cdot d}{\nu} \tag{3.65}$$

So Reynolds's criterion is: The flow is laminar if Re is small enough. The flow tends to be turbulent if Re is sufficiently large. The range in between is the so-called transition zone.

How could Reynolds tell turbulent from laminar? For the laminar case, he colored the center stream line with dye. When he increased the velocity of the fluid inside the pipe, the line became wavy (transition), and finally the dye was completely dispersed all over the pipe (fully developed turbulence).

A first explanation may be this: "Turbulence has to do with *instability*. The higher Re, the more unstable is the flow and the larger is the amplification of small perturbations."

If an experimenter repeats an experiment he tries to keep the initial conditions consistent. However, he can do this only to a certain extent. Small deviations *always* remain. Thus, if the experiment is sensitive to these tiny deviations, the result can be quite different. This is how turbulence is considered from the point of view of deterministic chaos.

However, if a turbulent experiment is repeated many times, it is seen that the flows are very similar *on average*. In statistics, there are many ways to take an average. If experiments are repeated N times, the so-called "ensemble average" is taken:

$$\overline{\Psi} = \lim_{N\to\infty}\left(\frac{1}{N}\sum_{n=1}^{N}\Psi_n\right) \approx \frac{1}{N}\sum_{n=1}^{N}\Psi_n \quad \text{for large } N \tag{3.66}$$

In equation (3.66), Ψ can be any quantity measured at a certain location at a distinct time. Figure 3.3 depicts the velocity component w measured at a fixed location in the course of time for a turbulent flow. The grey line indicates the (ensemble) average. This similarity on average is the motivation for a statistical viewpoint which is further pursued in the next section.

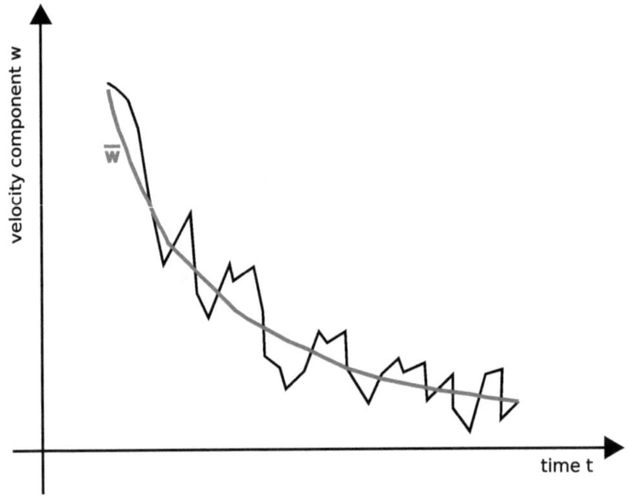

Figure 3.3: Average of the velocity component w.

What has been said above is by far not the whole story. A short and more sophisticated definition is given by Tsinober [137]: *"Turbulence is the manifestation of the spatiotemporal chaotic behavior of fluid flows at large Reynolds numbers, i. e., of a strongly nonlinear dissipative system with an extremely large number of degrees of freedom described by the Navier–Stokes equations."* Consult the literature for further explanations on the nature of turbulence [35, 110].

3.4.2 The k-ε model

The k-ε model [84] is the most frequently employed CFD turbulence model. It has the advantage of being computationally cheap and sufficiently accurate for many problems in engineering. It shall be described in what follows since it has been used in this work.

Technically speaking, the k-ε model is a two-equation "Reynolds Averaged Navier–Stokes equations" (RANS) turbulence model that makes use of the eddy viscosity hypothesis. Let us remember that turbulent flows are similar on average in order to understand the last sentence. As we cannot predict turbulent flows in detail, we hope it is possible to find an equation for the average. The starting point of Reynolds was to average the Navier–Stokes equations. (We shall use the ensemble average, equation (3.66), since it can be regarded as the most general average being also applicable in the case of temporal and spatial variations.) To appreciate the RANS model, note that from a statistical point of view, every flow quantity, Ψ, can be decomposed into an average, $\overline{\Psi}$, and a fluctuating part, Ψ':

$$\Psi = \overline{\Psi} + \Psi' \quad \text{Reynolds decomposition} \tag{3.67}$$

It is assumed that the flow is incompressible. Thus, from averaging the continuity equation it follows [110]

$$\nabla \boldsymbol{U} = 0 \quad \Rightarrow \quad \nabla \overline{\boldsymbol{U}} = 0 \quad \text{and} \quad \nabla \boldsymbol{U}' = 0 \tag{3.68}$$

The averaging process is also applied to the momentum equations (in nonconservation form), i. e.,

$$\frac{D\boldsymbol{U}}{Dt} = -\frac{1}{\rho}\nabla p + \nu\nabla^2 \boldsymbol{U} \tag{3.69}$$

The result is the RANS equations [110]:

$$\frac{\bar{D}\overline{U_j}}{\bar{D}t} = \frac{\partial \overline{U_j}}{\partial t} + \overline{\boldsymbol{U}} \cdot \nabla \overline{U_j} = \nu\nabla^2 \overline{U_j} - \frac{1}{\rho}\frac{\partial \overline{p}}{\partial x_j} - \frac{\partial \overline{U_i' U_j'}}{\partial x_i} \tag{3.70}$$

Equation (3.70) is written in tensor notation with repeated indices implying summation. Note that the differential operator $\bar{D}/\bar{D}t$ is defined by

$$\frac{\bar{D}}{\bar{D}t} = \frac{\partial}{\partial t} + (\overline{\boldsymbol{U}} \cdot \nabla). \tag{3.71}$$

(It is the temporal change of a quantity in an infinitesimal fluid element moving with the averaged velocity.) We see that in the RANS equations only averaged quantities appear *except* for the *Reynolds stresses*, $\overline{U_i' U_j'}$. At present, no method is known to predict the Reynolds' stresses from first principles. They have to be modeled.

One way to model the Reynolds' stresses is the so-called eddy viscosity hypothesis. Before we can state it, we need to define k, the turbulent kinetic energy per unit mass of the fluctuating field:

$$k \equiv \frac{1}{2}\overline{\boldsymbol{U}' \cdot \boldsymbol{U}'} = \frac{1}{2}\overline{U_i' U_i'} \tag{3.72}$$

The eddy viscosity hypothesis can be regarded as the mathematical analogue of the modeling of Newtonian fluids, equations (3.20). It defines the eddy viscosity, ν_t, also called turbulent viscosity:

$$-\overline{U_i'U_j'} + \frac{2}{3}k\delta_{ij} = \nu_t\left(\frac{\partial \overline{U}_i}{\partial x_j} + \frac{\partial \overline{U}_j}{\partial x_i}\right) \tag{3.73}$$

In equation (3.73), δ_{ij} is the Kronecker delta, which is either one for $i = j$ or zero otherwise. It is employed to keep equation (3.73) correct if a summation is implied by equal indices. Together with the kinematic viscosity, ν (fluid property), the turbulent viscosity, ν_t (flow property), contributes to an effective viscosity

$$\nu_{\text{eff}} = \nu + \nu_t, \tag{3.74}$$

which is put into the RANS equations to give

$$\frac{\bar{D}\overline{U}_j}{\bar{D}t} = \frac{\partial}{\partial x_i}\left[\nu_{\text{eff}}\left(\frac{\partial \overline{U}_i}{\partial x_j} + \frac{\partial \overline{U}_j}{\partial x_i}\right)\right] - \frac{1}{\rho}\frac{\partial}{\partial x_j}\left(\overline{p} + \frac{2}{3}\rho k\right) \tag{3.75}$$

The first question that arises is, "How large is ν_t?" Again, this has to be modeled. The answer given by the k-ε model is

$$\boxed{\nu_t = C_\mu k^2/\varepsilon}, \tag{3.76}$$

where C_μ is a model constant, usually set equal to 0.09, and ε is the *dissipation of turbulent kinetic energy*:

$$\varepsilon \equiv 2\nu\overline{s_{ij}s_{ij}} \quad \text{with } s_{ij} = \frac{1}{2}\left(\frac{\partial U_i'}{\partial x_j} + \frac{\partial U_j'}{\partial x_i}\right) \tag{3.77}$$

Provided the initial and boundary conditions for k and ε are known, their time profile is prescribed by two equations. The development of k is controlled by

$$\frac{\bar{D}k}{\bar{D}t} = \nabla\cdot\left(\left(\nu + \frac{\nu_t}{\sigma_k}\right)\nabla k\right) + \mathcal{P} - \varepsilon \tag{3.78}$$

In equation, (3.78) σ_k is a model parameter, usually set equal to one. The symbol $\mathcal{P}$ denotes the production of turbulent kinetic energy calculated by

$$\mathcal{P} \equiv -\overline{U_i'U_j'}\frac{\partial \overline{U}_i}{\partial x_j} \tag{3.79}$$

Equation (3.78) is a balance equation. The first term on the right-hand side stands for many others that are unknown.

The evolution of the dissipation is modeled in a similar way:

$$\frac{\bar{D}\varepsilon}{\bar{D}t} = \nabla \cdot \left(\left(\nu + \frac{\nu_t}{\sigma_\varepsilon} \right) \nabla \varepsilon \right) + C_{\varepsilon 1} \frac{\mathcal{P}\varepsilon}{k} - C_{\varepsilon 2} \frac{\varepsilon^2}{k} \tag{3.80}$$

Equation (3.80) can be interpreted as follows: Rate of change of ε plus the transport of ε by convection is equal to the transport by diffusion plus the rate of production of ε minus the rate of destruction of ε. The remaining model parameters are usually given the following values: $\sigma_\varepsilon = 1.3$, $C_{\varepsilon 1} = 1.44$, and $C_{\varepsilon 2} = 1.92$.

Davidson [35] pointed out three reasons why the eddy viscosity hypothesis is of limited use:

1. The Reynolds stress is related to the mean strain rate by a constant, ν_t, and not a tensor. This is not valid for strongly anisotropic turbulence.
2. If the mean strain rate is zero (e. g., for uniform mean flow), then $\overline{(u')^2} = \overline{(v')^2} = \overline{(w')^2}$ which is not always true.
3. The Reynolds' stress is controlled by the *local* rate of strain in the mean flow, not by the *history* of the straining of the turbulence. This may lead to wrong results if there is rapid straining by the mean flow.

Pope [110] brought together many tests of various turbulence models. On the evaluation of the k-ε model, he concluded:

- The k-ε model performs well for two-dimensional thin shear flows provided that the mean streamline curvature and the mean pressure gradient are small.
- For flows where the shear is highly complex (e. g., impinging jet, 3D flows), it fails profoundly.
- No secondary flows can be calculated due to the isotropic ν_t.
- The k-ε model is not suited for flows with strong swirl or mean rotation and flows with rapid variations in the mean flow.
- The dissipation equation can be improved by adjusting the constants. A general improvement applying to all cases has not been found.

There are other numerical approaches like LES and DNS that can simulate turbulence in a much more accurate way. However, they have not been employed for this monograph due to their immense computational costs. Therefore, they shall be mentioned only shortly here.

DNS means "Direct Numerical Simulation." It makes use of the fact that turbulence, in principle, is also governed by the Navier–Stokes equations like the laminar and transitional flows. However, the spatial and temporal resolution must be high enough to take into account the smallest scales. Since turbulence is three-dimensional, the number of cells, N_{cells}, is very large, and tends to be at the limit of the system. For a DNS, the Reynolds number is defined by

$$\text{Re} = \frac{U\ell}{\nu}, \tag{3.81}$$

where U is a typical velocity and ℓ is the size of the large, energy-containing eddies, the so-called integral scale. Davidson [35] shows that the maximal Re depends on the integral scale, on the size L_{BOX} of the control volume, and on N_{cells}:

$$\mathrm{Re} \sim \left(\frac{\ell}{L_{\mathrm{BOX}}}\right)^{4/3} N_{\mathrm{cells}}^{4/9} \tag{3.82}$$

The number of time steps is approximately given by

$$N_t \sim \frac{t_{\mathrm{total}}}{\ell/U} \mathrm{Re}^{3/4}, \tag{3.83}$$

where t_{total} is the total simulated time. Thus, the following relation for the computer time is gained:

$$\text{computer time} \propto N_{\mathrm{cells}} N_t \sim \left(\frac{t_{\mathrm{total}}}{\ell/U}\right)\left(\frac{L_{\mathrm{BOX}}}{\ell}\right)^3 \mathrm{Re}^3 \tag{3.84}$$

It is this law that prevents DNS from simulating large Reynolds' numbers despite the fast progress in computer technology.

LES is the abbreviation of "Large Eddy Simulation." For such a calculation, the smallest scales are not resolved but rather modeled by a so-called "sub-grid model." Typically, an order of magnitude is gained for the cell size. More recent monographs on LES are, e. g., that by Lesieur et al. [87] and that by Fröhlich [51] (in German).

Since we are interested in falling films, the influences of the wall and of the free surface has to be taken into account. This shall be done in the subsequent sections.

3.4.3 Turbulence near a wall

The velocity close to the wall is very low on account of the no-slip condition. Although the bulk flow may be turbulent, in the vicinity of the wall it is partially laminar (not completely). This region is called "the viscous sub-layer" since the viscous stress outweighs the Reynolds' stress. There, the dimensionless velocity, U^+, is connected to the dimensionless distance to the wall, y^+, by

$$U^+ = y^+, \tag{3.85}$$

where U^+ is defined as

$$U^+ = \overline{U}/U_\tau \quad \text{with} \quad U_\tau = \sqrt{\tau_w/\rho} \tag{3.86}$$

and y^+ is calculated from

$$y^+ = U_\tau y/\nu \tag{3.87}$$

In the second part of equation (3.86), τ_w denotes the shear stress at the wall. Expressed in terms of y^+, the viscous sub-layer is referred to as the region with $y^+ < 5$.

The region near to the viscous sub-layer ($30 < y^+ < 500$) is influenced by the solid boundary as well as by the turbulent bulk flow. Here, the so-called "log-law of the wall" governs the flow:

$$U^+ = \frac{1}{\kappa} \ln y^+ + A \tag{3.88}$$

In equation (3.88), κ denotes von Karman's constant the empirical value of which is about $\kappa \approx 0.4$. For a smooth wall, the additive constant A has been measured to be approximately 5.5. The roughness height also has to be considered for a rough wall; cf., e. g., Davidson [35].

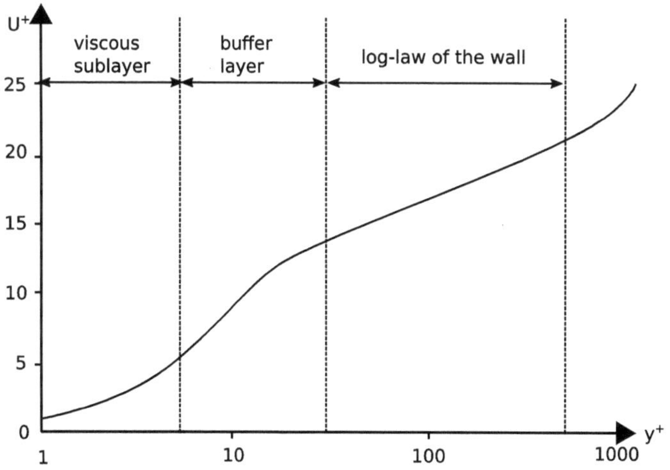

Figure 3.4: Relation between dimensionless velocity and dimensionless wall distance in case of a smooth wall.

Figure 3.4 shows the relation between U^+ and y^+ in the viscous sub-layer, in the buffer layer, and the log-law region in the case of a smooth wall.

In CFD, the influence of the wall on the turbulent flow is modeled by so-called "wall functions." It is assumed that the cell nearest to the wall is situated in the log-law region in order to apply these functions. However, for this book wall functions could not be employed since the thickness of the boundary layer exceeded the film thickness. Instead the low Re k-ε model of Lien, Chen, and Leschziner [88] was used which shall be outlined in the remainder of this section.

In order to account for the damping of the turbulence in the vicinity of the wall, a transition function f_μ is introduced in equation (3.76):

$$\nu_t = f_\mu \cdot C_\mu k^2 / \varepsilon \tag{3.89}$$

In this model, the transition function is

$$f_\mu = [1 - \exp(-0.02\mathrm{Re}_y)] \cdot \left(1 + \frac{5.3}{\mathrm{Re}_y}\right) \quad \text{with } \mathrm{Re}_y = \frac{y\sqrt{k}}{\nu}. \tag{3.90}$$

The equation for the turbulent kinetic energy k, equation (3.78), remains unchanged, whereas equation (3.80) for the dissipation rate ε is modified:

$$\frac{\bar{D}\varepsilon}{\bar{D}t} = \nabla \cdot \left(\left(\nu + \frac{\nu_t}{\sigma_\varepsilon}\right)\nabla\varepsilon\right) + f_1 \cdot C_{\varepsilon 1}\frac{\mathcal{P}\varepsilon}{k} - f_2 \cdot C_{\varepsilon 2}\frac{\varepsilon^2}{k} \tag{3.91}$$

The new functions, f_1 and f_2, are calculated by

$$f_1 = 2.33 \cdot [1 - 0.3\exp(-\mathrm{Re}_t^2)] \cdot \left[1 + 2\frac{\nu k}{y^2}\left(\overline{U_i' U_j'}\frac{\partial \overline{U_i}}{\partial x_j}\right)^{-1}\right] \cdot \exp(-0.00375\mathrm{Re}_y^2) \tag{3.92}$$

and

$$f_2 = 1 - 0.3\exp(-\mathrm{Re}_t^2). \tag{3.93}$$

The turbulent Reynolds' number, Re_t, is defined by

$$\mathrm{Re}_t = \frac{k^2}{\nu \cdot \varepsilon}. \tag{3.94}$$

Near the wall, the numerical mesh has to be refined such that the centroid C of the nearest cell has a distance $y^+ \sim 1$. The dissipation rate ε has a steep rise near the wall, but has to be zero at the wall surface. This decline is not resolved in the low Re model. Instead, at C, the value of the dissipation rate is calculated from [29]

$$\varepsilon_C = \mathcal{P} + \frac{2\nu k}{y^2}. \tag{3.95}$$

With that, the influence of the wall is considered.

3.4.4 Turbulence near a free surface

Still today the turbulence near a gas-liquid interface is not sufficiently understood. There are three attitudes of modeling the eddy viscosity, ν_t, inside a falling film which are outlined in Figure 3.5. In this figure, y denotes the distance from the wall and δ is the film thickness. Thus $y/\delta = 1$ is the location of the gas-liquid interface. Limberg [89] assumed no influence of the interface on the turbulence at all, whereas Mills and Chung [97] predicted a total suppression of the turbulence there. Mitrovic [98] suggested a kind of compromise between these extremes: The surface tension damps the

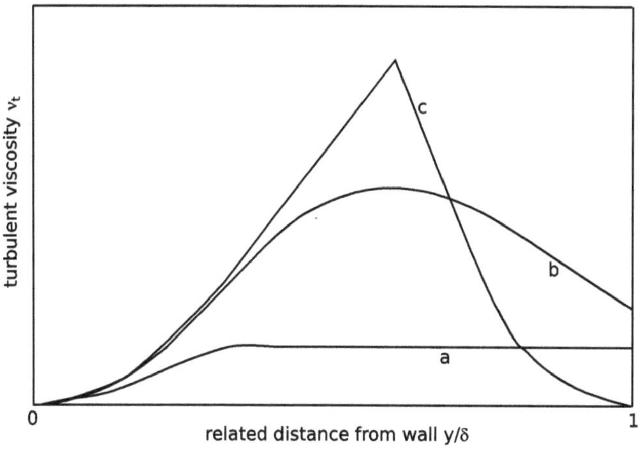

Figure 3.5: Three different ways to model the turbulence inside a falling film: (a) The gas-liquid interface has no influence. (b) The turbulence is damped there. (c) The turbulence vanishes at the film surface.

turbulence near the interface. However, this damping is not maximal. It only reduces the eddy viscosity to a certain amount. To be specific, for the mixing length ℓ, Mitrovic made the following ansatz:

$$\ell = 0.4y\left(1 - \exp\left(\frac{y^+}{26}\right)\right) \cdot \exp\left(-1.116\left(\frac{\sigma}{\delta\tau_c}\right)^{0.275}\left(\frac{y}{\delta}\right)^3\right) \tag{3.96}$$

with

$$\tau_c = 0.5(\tau_w + \tau_i). \tag{3.97}$$

In equation (3.96), σ is the surface tension. In equation (3.97), the subscript "i" refers to the interface. In the case $\tau/\tau_w = 1$, the eddy viscosity, ν_t, is obtained from

$$\frac{\nu_t}{\nu_L} = \frac{2 \cdot (\ell^+)^2}{1 + \sqrt{1 + 4 \cdot (\ell^+)^2}} \quad \text{where } \ell^+ = \frac{\ell \cdot U_\tau}{\nu_L} \tag{3.98}$$

According to Mitrovic, the distance from the wall, the film thickness, the surface tension, and the shear stress at the interface have an influence on the turbulence near the surface. Other models shall be briefly outlined in the following:

Lemos [86] follows Harlow and Nakayama by substituting the eddy viscosity, ν_t, by

$$\nu_t \rightarrow \nu_t \cdot \frac{1 - \exp(-\beta\nu/\nu_t)}{\beta\nu/\nu_t}, \tag{3.99}$$

i. e., the effective eddy viscosity is reduced for low-intensity turbulence. For the empirical constant β, a value of 100 is recommended.

Hagiwara and Madarame [62] notice that the “consumed” surface energy is

$$E_{\text{surf}} = \frac{\sigma}{r}, \tag{3.100}$$

where r denotes the curvature radius of disturbance. Thus, they propose a reduction of the turbulent kinetic energy, k, at the surface

$$k_{\text{surf}} = k - \frac{2\sigma}{\xi}. \tag{3.101}$$

In equation (3.101), ξ is the scale of the large turbulent eddies. A constraint is imposed on k_{surf}:

$$\frac{1}{2}(\overline{u'u'} + \overline{v'v'}) \leq k_{\text{surf}} \tag{3.102}$$

The free surface normal coincides with the z axis in the setup of Hagiwara and Madarame.

Lorencez et al. [92] state that k_{surf} is composed of streamwise and vertical fluctuations:

$$k_{\text{surf}} = k_u + k_v \tag{3.103}$$

The streamwise part, k_u, is considered proportional to the interfacial shear stress, which acts as a promoter of the tangential turbulence.

$$k_u = \frac{|\tau_i|}{\rho_L} \tag{3.104}$$

The contribution from the vertical fluctuations is modeled by

$$k_v = \left(\frac{2\Delta H}{t_{\text{period}}}\right)^2, \tag{3.105}$$

where ΔH is the mean (root mean square) wave amplitude and t_{period} the wave period both specified by experiment. The value for the dissipation rate, ε, is calculated from

$$\varepsilon_{\text{surf}} = \frac{k_{\text{surf}}^{3/2}}{a \cdot H_L} \tag{3.106}$$

In equation (3.106), H_L is the mean liquid height. The constant a is set equal to 0.2 for flows with a wavy interface.

Ferreira et al. [49] are cited to give a more recent reference. They impose

$$\frac{\partial k}{\partial n} = 0 \quad \text{and} \quad \frac{\partial \varepsilon}{\partial n} = 0, \tag{3.107}$$

i. e., they treat the free surface as a symmetry plane for k and ε, where no change is to occur.

3.4.5 Extension of the k-epsilon model for a free surface

In the following, a new turbulence model [115] is described which is an extension of the k-ε model. This extension is based an dimensional analysis, a widely used approach in turbulence research.

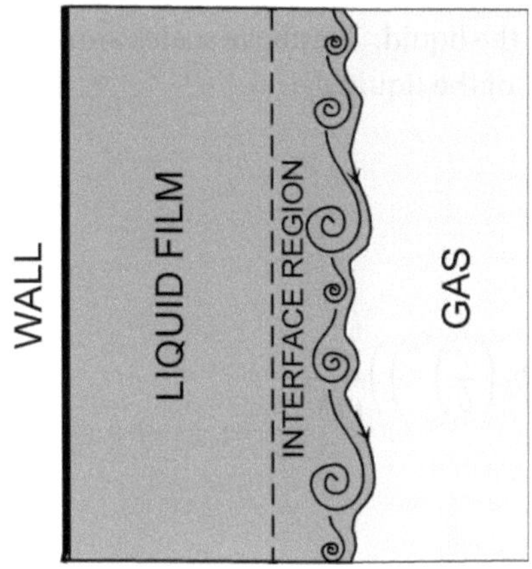

Figure 3.6: Eddies at the surface of a liquid film.

In a turbulent falling film across a wall, a multitude of eddies are to be expected in the vicinity of the liquid-gas interface, Figure 3.6. The large eddies transport the turbulent kinetic energy k, whereas the small eddies decay by viscous effects and dominate the dissipation rate ε. The basic idea of the model assumes that the large eddies decay by the action of the surface tension along a cascade with smaller and smaller eddies. The kinetic energy of the smallest eddies is dissipated into thermal energy at the end of the cascade. This picture is an extension of the Richardson hypothesis of the turbulence cascade in connection with viscosity.

Essentially, the k-ε model consists of two equations for the substantial derivatives of k and ε. The (averaged) substantial derivative is the rate of change of a quantity inside an infinitesimal volume that moves with the (mean) fluid flow. Because of a damping, a sink S_k should be introduced in the equation for $\bar{D}k/\bar{D}t$ in the vicinity of the free surface, and, consequently, a source S_ε in the equation for $\bar{D}\varepsilon/\bar{D}t$. These new terms should account for the surface tension σ, for the higher σ is, the stronger is the damping. Furthermore, these terms should only affect the liquid flow, inside the gas phase they are supposed to vanish. The "Volume of Fluid" scalar c_{VOF} can be employed for this purpose which in the case of a liquid-gas system is the volume fraction of the liquid in a cell of the numerical grid; cf. Chapter 5 and [65]. For instance, c_{VOF} is equal to 1 in a cell completely filled with liquid and is 0 in a cell where there is only gas.

Estimations for the time and length scales $(t_\xi, \xi, t_\eta, \eta)$ of the largest and smallest eddies based on dimensional analysis can be found in the literature [42]:

$$t_\xi = \frac{k}{\varepsilon}, \quad \xi = \frac{k^{3/2}}{\varepsilon} \quad \text{(for the largest eddies)} \tag{3.108}$$

$$t_\eta = \sqrt{\frac{\nu_L}{\varepsilon}}, \quad \eta = \left(\frac{\nu_L^3}{\varepsilon}\right)^{1/4} \quad \text{(for the smallest eddies)} \tag{3.109}$$

The symbol ν_L denotes the kinematic viscosity of the liquid. The above scales are combined with the surface tension σ and the density of the liquid ρ_L:

$$\begin{aligned}\frac{\bar{D}k}{\bar{D}t} &= \Sigma_k - S_k \\ &= \Sigma_k - C_{\sigma 1}\frac{\sigma/\xi}{\rho_L t_\xi} \\ &= \Sigma_k - c_{\mathrm{VOF}} a_k \left(1 - \exp\left(-b_k\left(\frac{y}{\delta}\right)^{c_k}\right)\right)\frac{\sigma\varepsilon^2}{\rho_L k^{5/2}}\end{aligned} \tag{3.110}$$

$$\begin{aligned}\frac{\bar{D}\varepsilon}{\bar{D}t} &= \Sigma_\varepsilon + S_\varepsilon \\ &= \Sigma_\varepsilon + C_{\sigma 2}\frac{\sigma/\eta}{\rho_L t_\eta^2} \\ &= \Sigma_\varepsilon + c_{\mathrm{VOF}} a_\varepsilon \left(1 - \exp\left(-b_\varepsilon\left(\frac{y}{\delta}\right)^{c_\varepsilon}\right)\right)\frac{\sigma\varepsilon^{5/4}}{\rho_L \nu_L^{7/4}}\end{aligned} \tag{3.111}$$

In equations (3.110) and (3.111), y is the distance from the wall and δ is the film thickness; Σ_k and Σ_ε denote the terms of the ordinary k-ε model. The expressions in brackets are supposed to be maximal in the vicinity of the free surface. In equation (3.110), only the time and length scales of the largest eddies are considered which have the dominant influence on the turbulent kinetic energy k. Analogously, in equation (3.111) there are only the scales of the smallest eddies, since on these scales the dissipation takes place.

In order to close the model, values have to be specified for the coefficients a_k, b_k, c_k, a_ε, b_ε, and c_ε. For this purpose, numerous numerical experiments would be necessary that determine the heat transfer coefficient which could then be compared to experiments (why the experiments have to be thermal, shall be explained at the end of this section). The accomplishment of this task is beyond the scope of this monograph. However, first simulations of the new model with Star-CD are reported in Chapter 6.

Note that for Σ_k and Σ_ε, also terms of a low Reynolds' number k-ε model can be used which is better suited to describe turbulent thin liquid films. Of course, any k-ε model may be extended like in equations (3.110) and (3.111).

In many commercial CFD programs, there is the possibility to add source terms in the equations for k and ε. Therefore, this model may be well suited for the engineer who employs commercial CFD software.

A short explanation, why thermal experiments are well suited to test the distribution of ν_t inside falling films, is given at the end of this section. For this purpose, consider the following two equations from Mudawwar and El-Masri [102]:

$$1 - \frac{y}{\delta} = \left(1 + \frac{\nu_t}{\nu_L}\right)\frac{dU^+}{dy^+} \tag{3.112}$$

$$\frac{q}{q_w} = \frac{1}{\mathrm{Pr}}\left(1 + \frac{\mathrm{Pr}}{\mathrm{Pr}_t} \cdot \frac{\nu_t}{\nu_L}\right)\frac{dT^+}{dy^+} \quad \text{with} \quad T^+ = \frac{\rho_L c_p U_\tau (T_w - T)}{q_w} \tag{3.113}$$

The symbol c_p denotes the specific heat at constant pressure. The Prandtl number, Pr, is defined as the ratio of the viscous diffusivity, ν, and the thermal diffusivity, $D_{\mathrm{th}} = \lambda/(\rho c_p)$, equation (3.34). Analogously, the turbulent Prandtl number, Pr_t, is the turbulent viscous diffusivity, ν_t, divided by the turbulent thermal diffusivity, $D_{\mathrm{th},t}$, defined by

$$-\overline{\boldsymbol{U}'T'} = D_{\mathrm{th},t}\nabla\overline{T} \tag{3.114}$$

Now, let us turn to the content of the equations (3.112) and (3.113): The first equation serves as a means to determine the film thickness after a distribution for ν_t has been specified. It is seen from equation (3.112) that much depends on the region near the wall, where $y/\delta \ll 1$. However, in the vicinity of the free surface, the left-hand side of equation (3.112) vanishes. Therefore, the distribution of ν_t in this region does not matter so much. The picture changes for equation (3.113): Here, the heat is transported from the wall to the surface for an evaporating film. Thus, the left-hand side of this equation, the ratio of the heat flux and the heat flux at the wall, is rather constant. The distribution of the eddy viscosity is also important to know near the liquid-gas interface for this reason.

4 Fundamentals of falling films

In this chapter, the foundations are explained in order to understand the numerical experiments in the Chapters 6, 7, 8, and 9. A lot can be said on falling films, especially on the mathematics of wavy, laminar films at small and moderate Reynolds' numbers. However, there would not be much use in repeating the previously published monographs [13, 30, 72]. Instead many mathematical details shall be omitted here. Usefulness for numerical applications is the most important criterion.

Figure 4.1 shows a wavy falling film. Tap water at approximately 20 °C flows down a vertical plate which is 290 mm high and 400 mm wide. The Reynolds' number, to be defined in just a moment, is in the order of 100. Near the distributor, the film has a smooth surface. In the region of the water tap (not to be seen on this picture), the first waves develop. Further downstream, more and more waves are spreading, eventually covering the whole width of the plate. Although the height is only 290 mm, it can already be seen that the degree of disorder is increasing in the downstream direction. Falling films like this one are our main concern.

In this chapter, the different flow regimes shall be explained, since the Reynolds' number has a strong influence on the falling film. The flow needs a certain distance from the inlet, before it can be classified as being developed. This entrance region shall be examined a bit closer. Then we want to understand under which circumstances instabilities do increase. This question of stability shall be understood rather intuitively from physical reasoning, complicated mathematical equations are to be avoided. Some typical flow patterns, observed in real experiments, are introduced. Furthermore, experimental correlations are an important tool to the engineer. Some

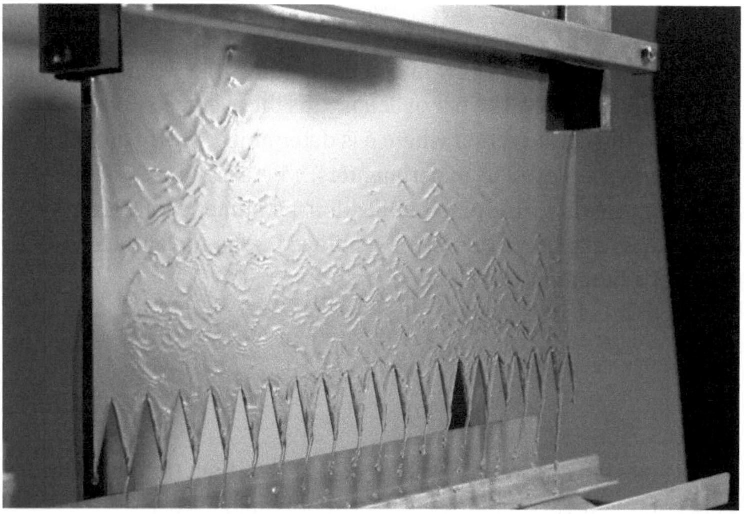

Figure 4.1: Undisturbed wavy falling film.

https://doi.org/10.1515/9783110592337-004

of them shall be mentioned in a separate section. Moreover, a two equation model for falling films shall be quoted which predicts the mean velocity and the film thickness. It is formulated in a dimensional form. A mathematician prefers the dimensionless version, whereas an engineer regards the dimensional form as more practical to do physical calculations. The two equations are named "long wave equations," for a long wavelength is assumed. Moreover, references on other simulations of falling films will be given. Separate sections deal with the zero streamline behind an obstacle, the mixing effect in evaporating turbulent saltwater films, and what reasonable approximations can be made.

Throughout the remaining chapters, the following coordinate system shall be used: The direction parallel to the wall is called *streamwise*. The streamwise distance from the inlet is the coordinate x. The thickness of the film is measured by the *cross-stream* coordinate y. For simulations in three dimensions, we need the *spanwise* coordinate z which is limited by the width of the plate. Vertical walls are our main concern. However, also inclined plates shall be considered from time to time. Therefore, we define the inclination angle θ which is 90° for a vertical wall.

4.1 Flow regimes

Figure 4.2 shows the various flow regimes of a falling film that may occur during condensation of vapor at a cooled vertical wall. Two numbers are of importance here: the Reynolds' number, Re, and the Kapitza number, Ka. (Ka is also often referred to as film number, Fi, [13].) In the case of a falling film, the Reynolds' number is defined by

$$\mathrm{Re} = \frac{\dot{M}_L}{Z \cdot \mu_L} = \frac{\overline{u} \cdot \delta}{\nu_L} \tag{4.1}$$

In equation (4.1), $\dot{M}_L$ denotes the mass flow rate of the liquid, Z the width of the film, and μ_L the liquid dynamic viscosity. The bar over u means the average x velocity component inside the film at the cross-section where δ is determined. The first relation in equation (4.1) is frequently employed by experimenters, whereas the second one is often more convenient for theorists. (In some publications, a numerical factor appears in the definition of Re, since another velocity is employed, e. g., the surface velocity or the phase velocity of a harmonic wave.)

The Kapitza number is calculated from

$$\mathrm{Ka} = \frac{\rho_L \cdot \sigma^3}{g \cdot \mu_L^4} \tag{4.2}$$

where σ is the surface tension and g the gravitational acceleration. (In the literature, the definition is not consistent. Sometimes the cube root is employed.)

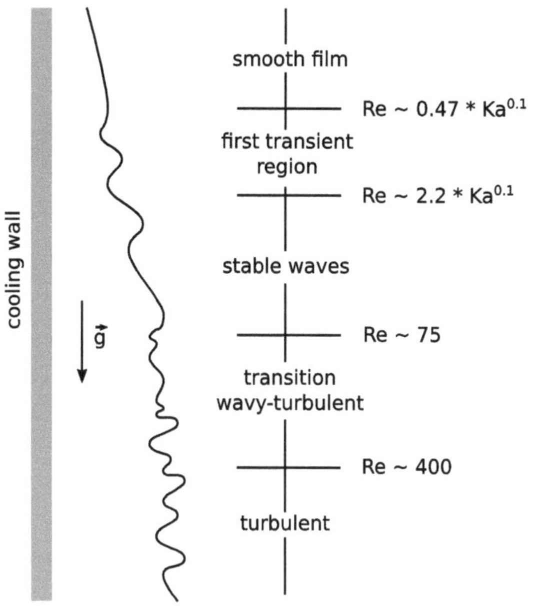

Figure 4.2: Different flow regimes for a condensing film.

The flow regimes of Figure 4.2 shall be explained in the following [13]:
1. $\mathrm{Re} \leq 0.47\mathrm{Ka}^{0.1}$, purely laminar, i. e., smooth film, for water at 20 °C the limit on Re is 5.4;
2. $0.47\mathrm{Ka}^{0.1} \leq \mathrm{Re} \leq 2.2\mathrm{Ka}^{0.1}$, first transient regime, instabilities begin to have a perceivable effect, the upper limit for water at 20 °C is 25;
3. $2.2\mathrm{Ka}^{0.1} \leq \mathrm{Re} \leq 75$, stable wavy flow;
4. $75 \leq \mathrm{Re} \leq 400$, second transient regime, first turbulent spots, transition to turbulence;
5. $\mathrm{Re} \geq 400$, fully turbulent flow.

A perfectly smooth film can be easily solved. Therefore, it will be given much attention in the next section.

4.2 The smooth film

"Nusselt's water skin" shall be treated in this section [107, 96]. Originally, the line of argument was for condensation. However, it is rather straightforward to make it valid for evaporation. The chapters on numerical experiments are largely founded on the theory of the smooth falling film. Therefore, calculations are presented here in detail.

Let us begin with the derivation of the velocity profile inside a smooth falling film. It is assumed that the film flow is stationary and unidirectional; cf. Figure 4.3. The force

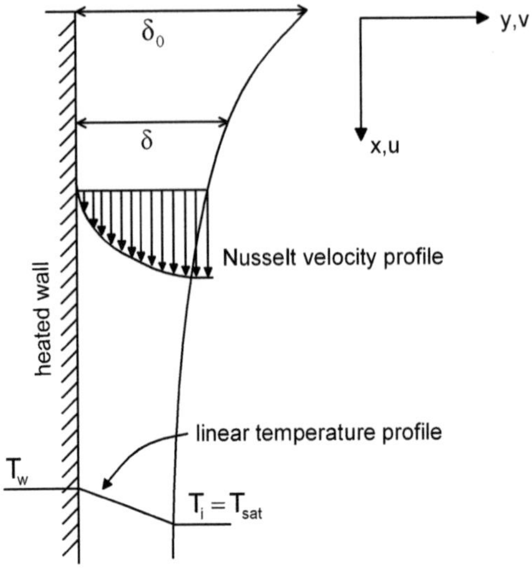

Figure 4.3: Evaporating smooth falling film.

balance for a fluid element inside the film is supposed to be

$$\rho_L\, g\, dx\, dyZ = dy\, Z(p_x - p_{x+dx}) + dx\, Z(\tau_{y+dy} - \tau_y) = 0 \tag{4.3}$$

Two expansions according to Taylor can be made and the higher terms omitted:

$$p_{x+dx} \approx p_x + \frac{\partial p_x}{\partial x} dx \tag{4.4}$$

$$\tau_{y+dy} \approx \tau_y + \frac{\partial \tau_y}{\partial y} dy \tag{4.5}$$

Now the indices x and y can be dropped, and equation (4.3) becomes

$$-\rho_L\, g\, dx\, dy + dy\, \frac{\partial p}{\partial x}\, dx - dx\, \frac{\partial \tau}{\partial y}\, dy = 0 \tag{4.6}$$

This gives

$$\frac{\partial \tau}{\partial y} = -\rho_L g + \frac{\partial p}{\partial x} \tag{4.7}$$

Since the hydrostatic pressure of the liquid has already been accounted for, the hydrostatic pressure rise in the gas shall also be considered by setting

$$\frac{\partial p}{\partial x} = \rho_G g. \tag{4.8}$$

The liquid is assumed to be Newtonian and the flow unidirectional:

$$\tau = \mu_L \frac{\partial u}{\partial y} \tag{4.9}$$

Thus, a differential equation is obtained:

$$\frac{\partial^2 u}{\partial y^2} = -\frac{1}{\mu_L}(\rho_L - \rho_G)g \tag{4.10}$$

Integration gives

$$u = -\frac{(\rho_L - \rho_G)g}{\mu_L}\frac{y^2}{2} + C_1 y + C_2 \tag{4.11}$$

The integration constants C_1 and C_2 follow from the boundary conditions

$$u = 0 \quad \text{at} \quad y = 0 \quad \text{(wall)} \quad \Rightarrow C_2 = 0 \tag{4.12}$$

$$\frac{\partial u}{\partial y} = \frac{\tau_i}{\mu_L} \quad \text{at} \quad y = \delta \quad \text{(interface)} \quad \Rightarrow C_1 = \frac{\tau_i}{\mu_L} + \frac{\rho_L - \rho_G}{\mu_L}g\delta \tag{4.13}$$

Finally, we obtain the velocity profile for a smooth falling film in the presence of a shear stress at the interface, τ_i:

$$u = \frac{\rho_L - \rho_G}{\mu_L}g\delta^2\left[\frac{y}{\delta} - \frac{1}{2}\left(\frac{y}{\delta}\right)^2\right] + \frac{\tau_i}{\mu_L}y \tag{4.14}$$

A negative interfacial shear slows the film down and results in a larger thickness. On the contrary, a positive shear accelerates the film diminishing δ. In the remainder of this section, a vanishing τ_i is assumed. Equation (4.14) becomes

$$\boxed{u = \frac{\rho_L - \rho_G}{\mu_L}g\delta^2\left[\frac{y}{\delta} - \frac{1}{2}\left(\frac{y}{\delta}\right)^2\right]} \tag{4.15}$$

It is this semi-parabolic velocity profile that shall be referred to as *the Nusselt velocity profile*. The maximal velocity is at the liquid-gas interface:

$$u_{\max} = u(\delta) = \frac{1}{2}\frac{\rho_L - \rho_G}{\mu_L}g\delta^2 \tag{4.16}$$

The average velocity is

$$\bar{u} = \frac{1}{\delta}\int_0^\delta u\,dy = \frac{1}{3}\frac{\rho_L - \rho_G}{\mu_L}g\delta^2 = \frac{2}{3}u_{\max} \tag{4.17}$$

(This kind of average and the ensemble average from turbulence research should not be confounded.) The mass flow rate is calculated from $\dot{M} = \bar{u}\rho_L Z\delta$ to give

$$\dot{M} = \frac{\rho_L(\rho_L - \rho_G)}{3\mu_L}gZ\delta^3 \tag{4.18}$$

Inserting equation (4.18) in the definition for the Reynolds' number, equation (4.1), and neglecting ρ_G in comparison to ρ_L, we obtain a correlation between the film thickness, δ, and Re:

$$\boxed{\delta = \left(3\,\mathrm{Re} \cdot \frac{\nu_L^2}{g}\right)^{1/3}} \tag{4.19}$$

Note that the expression $(\nu_L^2/g)^{1/3}$ has the dimension of a length.

After having dealt with the hydrodynamics, we now turn to the evaporating smooth falling film. The film is heated by the wall at constant temperature T_W. The heat is transported through the liquid film to the interface which is assumed to have the saturation temperature T_{sat}. The temperature profile inside the film is linear since the temperatures at the wall and at the interface are held constant. The liquid is regarded as a pure substance, the gas as pure vapor at constant temperature T_{sat}. In a narrow strip of extension dx, the heat flow rate $d\dot{Q}$ is correlated to the evaporated mass flow rate $d\dot{M}$ by the specific enthalpy difference $\Delta h = h_V - h_L$:

$$d\dot{Q} = \Delta h \, d\dot{M} \tag{4.20}$$

On account of the linear temperature profile inside the film, we have

$$d\dot{Q} = \lambda_L \frac{T_W - T_{\text{sat}}}{\delta} Z \, dx \tag{4.21}$$

From equation (4.18), we obtain

$$\frac{d\dot{M}}{d\delta} = \frac{\rho_L(\rho_L - \rho_G)}{\mu_L} gZ\delta^2 \tag{4.22}$$

From equations (4.20), (4.21), and (4.22), the following relation holds:

$$\lambda_L \frac{T_W - T_{\text{sat}}}{\delta} Z \, dx = \Delta h \frac{\rho_L(\rho_L - \rho_G)}{\mu_L} gZ\delta^2 \, d\delta \tag{4.23}$$

After a clean separation of the variables δ and x, this is equivalent to

$$\delta^3 \, d\delta = \frac{\lambda_L \mu_L (T_W - T_{\text{sat}})}{\rho_L(\rho_L - \rho_G) g \Delta h} dx \tag{4.24}$$

Provided that $\delta(x = 0) = \delta_0$, integration of equation (4.24) gives

$$\boxed{\delta = \left(\delta_0^4 - \frac{4\lambda_L \mu_L (T_W - T_{\text{sat}})}{\rho_L(\rho_L - \rho_G) g \Delta h} \cdot x\right)^{1/4}} \tag{4.25}$$

Now let us turn to the question how to evaluate the heat transfer through a falling film. For this purpose, we shall employ the *heat transfer coefficient*. The local heat transfer coefficient, α_x, can be understood as a "factor of proportionality" between the heat flux, q, and the driving temperature difference:

$$q = \frac{d\dot{Q}}{dA} = \frac{d\dot{Q}}{Z\,dx} = \alpha_x \cdot (T_w - T_{sat}) \tag{4.26}$$

From comparing equation (4.26) with equation (4.21), we obtain

$$\boxed{\alpha_x = \frac{\lambda_L}{\delta}} \tag{4.27}$$

From equation (4.27), it is obvious that *the thinner the film, the better is the heat transfer*. For a plate of height L, the mean heat transfer coefficient, $\overline{\alpha}$, is

$$\begin{aligned} \overline{\alpha} &= \frac{1}{L}\int_0^L \alpha_x\,dx \\ &= \frac{1}{L}\int_0^L \lambda_L(\delta_0^4 - 4Cx)^{-1/4}\,dx \\ &= \frac{\lambda_L}{3LC}[\delta_0^3 - (\delta_0^4 - 4CL)^{3/4}] \end{aligned} \tag{4.28}$$

From equation (4.25), it is seen that

$$C = \frac{\lambda_L \mu_L (T_w - T_{sat})}{\rho_L(\rho_L - \rho_G) g \Delta h}, \tag{4.29}$$

which is treated as a constant in equation (4.28). If a dimensionless number is required to specify the heat transfer, the *Nusselt number*, Nu, is employed:

$$\mathrm{Nu}_x = \frac{\alpha_x}{\lambda_L}\left(\frac{\nu_L^2}{g}\right)^{1/3} \quad \text{and} \quad \overline{\mathrm{Nu}} = \frac{\overline{\alpha}}{\lambda_L}\left(\frac{\nu_L^2}{g}\right)^{1/3} \tag{4.30}$$

Note that, in the definition of the Nusselt number for a falling film, $(\nu_L^2/g)^{1/3}$ is used as characteristic length. The local Nusselt number would be always one if δ was taken instead.

4.3 The entrance region

Coming from the distributor, the falling film needs some distance in order to develop hydrodynamically as well as thermally. Both aspects shall be treated in this section. However, it will be shown that the entrance region is rather short in comparison to the overall height of the plate which is confirmed by the numerical experiments in Chapter 6 and Chapter 9. Therefore, this topic is treated rather briefly here. Nonetheless, it may be important for plants with plates of small height.

4.3.1 Hydrodynamic point of view

Frequently, the velocity profile inside the distributor is assumed to be parabolic under reference to the Hagen–Poiseuille solution. Alekseenko et al. [13] go one step further and let the profile be semiparabolic everywhere which is depicted in Figure 4.4. The corresponding equation reads

$$u = \frac{g}{\nu_L} \frac{\delta_0^3}{\delta} \left[\frac{y}{\delta} - \frac{1}{2} \left(\frac{y}{\delta} \right)^2 \right], \tag{4.31}$$

where δ is the film thickness and δ_0 the thickness in equilibrium, i. e., the Nusselt solution according to equation (4.19). At the inlet, the film thickness has the value δ_{in}. Let us define

$$H_{in} = \frac{\delta_{in}}{\delta_0} \quad \text{and} \quad H = \frac{\delta}{\delta_0}. \tag{4.32}$$

The following equation from Alekseenko et al. [13] yields the distance x where H has a value between H_{in} and 1:

$$\begin{aligned} \frac{15x}{\mathrm{Re}\,\delta_0} &= 2\sqrt{3}\left[\arctan\left(\frac{2H+1}{\sqrt{3}}\right) - \arctan\left(\frac{2H_{in}+1}{\sqrt{3}}\right)\right] \\ &\quad - \ln \frac{(1-H)^2(1+H_{in}+H_{in}^2)}{(1-H_{in})^2(1+H+H^2)} \end{aligned} \tag{4.33}$$

For $H = H_{in}$ the right-hand side is zero, so is x on the left-hand side. For $H = 1$, the right-hand side yields infinity which means that the equilibrium can only be approached. Table 4.1 shows some values for x in dependence on H. The initial condition is $H_{in} = 4$.

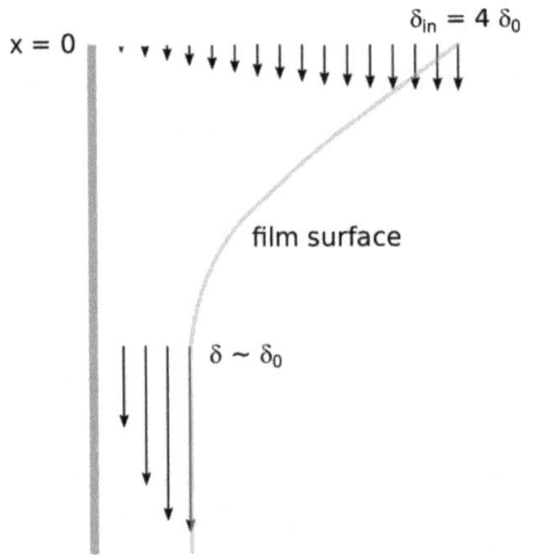

Figure 4.4: Modeling the velocity in the entrance region.

Table 4.1: The related distance from the inlet in dependence on the related film thickness.

δ/δ_0	4	3	2	1.5	1.3	1.1	1.05	1.01
x/δ_0	0	0.6	2.4	5.4	8.2	15.6	20.7	33.3

The Reynolds' number is 60. In order to calculate δ_0, a saltwater film at 60 °C is considered. Since in this case $\delta_0 = 0.167$ mm, it can be seen from the table that only after 6 mm the equilibrium is approached with a deviation less than 1 %.

From equation (4.31), we can estimate the wall shear stress, $\tau_w = \mu_L(\partial u/\partial y)_{y=0}$, and the surface velocity, $u_{surface} = u(y = \delta)$, in the entrance region:

$$\frac{\tau_W}{\rho_L g \delta_0} = \frac{1}{H^2} \tag{4.34}$$

$$\frac{u_{surface}}{u_{s0}} = \frac{1}{H} \quad \text{with } u_{s0} = \frac{g\delta_0^2}{2\nu_L} \tag{4.35}$$

So far, the hydrodynamics are taken into account.

4.3.2 Thermal point of view

There is also a development of the temperature field inside a falling film running down a heating wall. Figure 4.5 displays two cases. On the left-hand side, the thermal entrance region of a nonevaporating film is shown. The mean temperature is increasing until the whole film has the temperature of the wall, T_w. The wall temperature is assumed to be constant. In the $y - T$ diagram, the slope at the interface is approximately horizontal, which indicates that there is almost no heat transfer from the surface to the gas. On the right-hand side, the temperature inside an evaporating laminar film is displayed. It is assumed that the initial temperature is also the saturation temperature T_{sat}. The heat is transported from the wall to the surface where the temperature has a constant value $T_{surface} = T_{sat}$. Eventually, the temperature profile is linear. Its slope at the interface is no longer horizontal. This nonvanishing gradient causes the evaporation. The temperature profiles in the case of turbulence are not shown. In a certain distance from the wall, we would have turbulent mixing which has an equalizing effect on the temperature.

4.4 Stability

Under which conditions is a small perturbation amplified resulting in a wavy falling film? The answer to this question depends on the inclination angle, the flow rate, and the wavelength of the perturbation. In Figure 4.6, a falling film is depicted that is flowing down an inclined plane. In this picture, the inclination angle θ has a value of 45°.

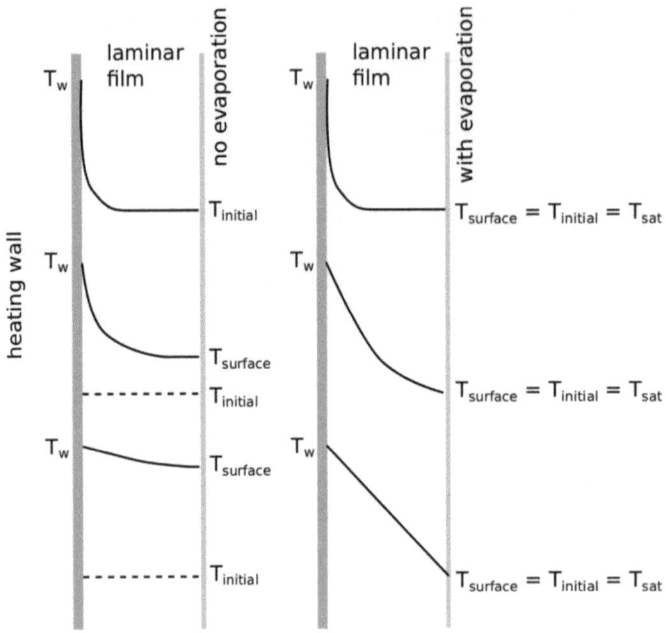

Figure 4.5: Developing temperature profile in the entrance region in case of a laminar film without (left) and with evaporation (right) at the surface.

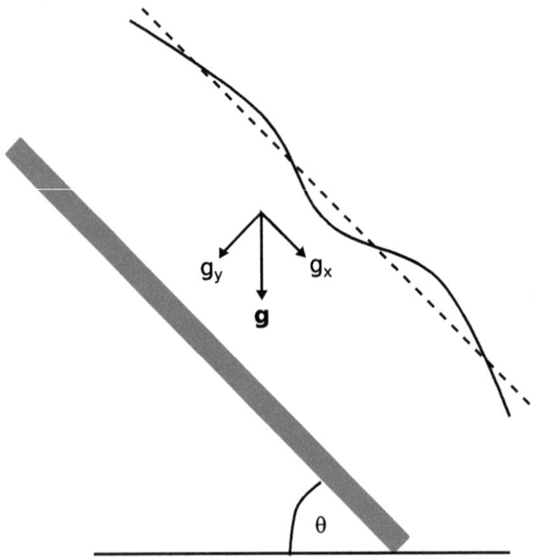

Figure 4.6: Falling film with perturbation running down an inclined plane.

The dashed line marks a smooth surface. There exists a small perturbation, a wave with crests and troughs. In this figure, the gravitational acceleration $\boldsymbol{g}$ is decomposed in a streamwise component g_x and a cross-stream component g_y. It is g_y that has a balancing effect. Let us consider a level inside the film with the distance from the wall y. From hydrostatics, we obtain the balancing pressure:

$$p_{\text{balancing}} = \rho_L g_y(\delta - y) \tag{4.36}$$

The larger the film thickness δ is, the larger is $p_{\text{balancing}}$. Thus under a wave crest the pressure is larger than under a neighboring trough. This causes a relative flow acting against the deformation of the surface.

This qualitative argument is confirmed in a quantitative stability analysis, not to be performed here. A small sinusoidal perturbation is imposed on a smooth film. The sinusoidal wave has a wavelength λ_x. This perturbation is only amplified if λ_x is large enough. The smaller θ is, the larger the wavelength has to be. So the result of the stability analysis can be summarized as, "*A falling film flowing down an inclined plane is unstable to long wave perturbations.*"

Again qualitatively, we can understand this dependence on the inclination angle θ. The balancing pressure, equation (4.36), depends on g_y which in turn depends on $\sin\theta$. Let us consider the case of an almost horizontal wall, $\theta \approx 0°$ and $g_y \approx g$. The driving gravitational component g_x for the film to fall would be very weak. The balancing pressure would be large. We would have to increase the Reynolds' number in order to overcome its stabilizing effect. The quantitative stability analysis confirms this reasoning. As a condition for the film to be unstable, it yields

$$\boxed{\mathrm{Re} > \cot\theta}\,. \tag{4.37}$$

Now let us consider the case $\theta = 90°$ which is of special interest for us. If the wall is vertical, the balancing pressure vanishes, since $g_y = 0$. Equation (4.37) becomes

$$\boxed{\mathrm{Re} > 0 \qquad \text{in case of a vertical wall}}\,. \tag{4.38}$$

The above condition means that *vertically falling films are unstable at every Reynolds' number* and waves should always exist. This statement seems to be contradicted by the observation that at very low Reynolds' numbers falling films are observed to have a smooth surface. This problem is solved if we take a look on the amplification over the time interval required to travel 100 times the film thickness. The amplification factor is [48]

$$f_{\text{amplification}} = \frac{|\delta(t_{100})|}{|\delta(t=0)|} = \exp\left\{\left[12.5\nu_L^{4/3} g^{1/3}\frac{\rho_L}{\sigma}\right]\mathrm{Re}^{8/3}\right\} \tag{4.39}$$

(In [48], the equation looks a bit different, since the Reynolds' number is defined in another way.) For saltwater at 60 °C, equation (4.39) becomes

$$f_{\text{amplification}} = \exp\{1.636 \times 10^{-3}\mathrm{Re}^{8/3}\} \tag{4.40}$$

Table 4.2: Amplification factor for different Reynolds' numbers.

Re	1	2	5	10	15	20	30
$f_{\text{amplification}}$	1.002	1.01	1.13	2.14	9.4	124	1.5×10^6

The salinity is $35\,\text{kg/m}^3$, the viscosity $\nu_L = 5.04 \times 10^{-7}\,\text{m}^2/\text{s}$, the density $\rho_L = 1008.7\,\text{kg/m}^3$, and the surface tension $\sigma = 0.066\,\text{N/m}$. In Table 4.2, the amplification factor is calculated for different Reynolds' numbers according to equation (4.40).

It can be seen from the table that at lower Reynolds' numbers the amplification of a wave is very weak, so a wave may not be perceived. From a certain value for Re, the amplification factor starts to be significantly larger than 1. Thus the observer gets the impression that at low Re the film is smooth and at larger values the falling film becomes wavy.

Why does the modulus of the film thickness appear in equation (4.39)? Is δ not always positive? The point is that in stability analysis δ is complex. The Nusselt solution, denoted by δ_0, is disturbed by δ':

$$\delta = \delta_0 + \delta' \tag{4.41}$$

This perturbation is assumed to be complex:

$$\delta' = \delta^* \exp\{ik_x(x - u_{\text{ph}}t)\} \tag{4.42}$$

In this equation, δ^* is the amplitude of the perturbation, $k_x = 2\pi/\lambda_x$ the wave number, and $u_{\text{ph}} = \lambda_x/t_{\text{period}}$ the phase velocity of the wave.

The time interval t_{100}, required to travel 100 times the film thickness, can be specified if we take into account that the velocity of a harmonic wave is

$$u_{\text{ph}} = 3\overline{u}_0 = \frac{\rho_L}{\mu_L} g\delta_0^2, \tag{4.43}$$

where $\overline{u}_0$ is the average velocity of a smooth film. In equation (4.43), equation (4.17) is used neglecting the density of the gas in comparison to the density of the liquid. Therefore,

$$t_{100} = \frac{100\delta_0}{u_{\text{ph}}} = \frac{100\nu_L}{g\delta_0}. \tag{4.44}$$

It has been mentioned that there is a critical wavelength, λ_c, below which the film remains flat. Beside of Re and θ, it also depends on the *Weber number*:

$$\text{We} = \frac{\sigma}{\rho_L \delta_0^2 g \sin\theta} \tag{4.45}$$

Only a relation, not an equation, is given here:

$$\lambda_c \propto \sqrt{\frac{\text{We}}{\frac{6}{5}\text{Re} - \cot\theta}} \tag{4.46}$$

We see that there is a lower value of the critical Reynolds' number

$$\mathrm{Re}_c = \frac{5}{6} \cot\theta. \tag{4.47}$$

This limit is a bit lower than in equation (4.37). It is the result of solving the Orr–Sommerfeld eigenvalue problem [72].

4.5 Flow patterns

Flow patterns of falling liquid films are a broad field from which only a few examples are mentioned in this section. To get a first impression, look at Figures 4.1 and 7.4. In Figure 4.1, a photograph of a falling film is displayed. Figure 7.4 shows the result of a three-dimensional simulation. Both figures have in common that the film is initially smooth, the waves appear further downstream. In Figure 7.4, the waves are first two-dimensional. Having a closer look on the crests and troughs at the bottom of this plot, we can perceive deviations from two-dimensionality. In Figure 4.7, the development from smooth to three-dimensionality is sketched. The surface of the falling film becomes more and more disordered. At the bottom of the drawing, horseshoe-shaped

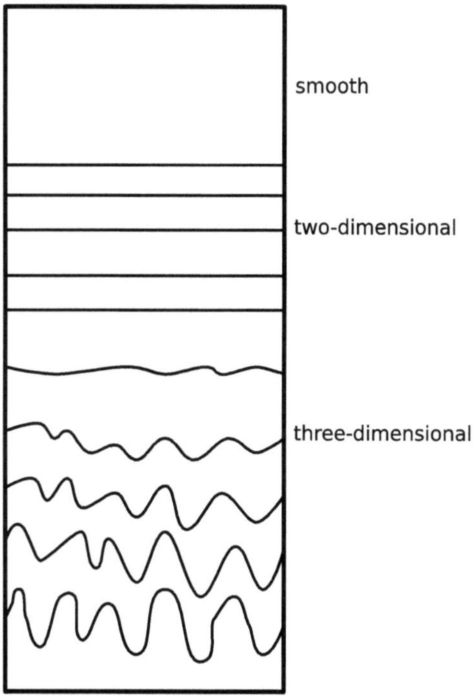

Figure 4.7: An initially smooth film becomes wavy. First, the waves are two-dimensional. Further downstream they are three-dimensional.

structures are indicated. In Figure 4.8, a typical pattern of a two-dimensional falling film is plotted. The example is taken from Annex B. The large waves are called *solitary waves* or *roll waves*. In front of them, there are the smaller *capillary waves*. Solitary and capillary waves will be further examined in Chapter 7.

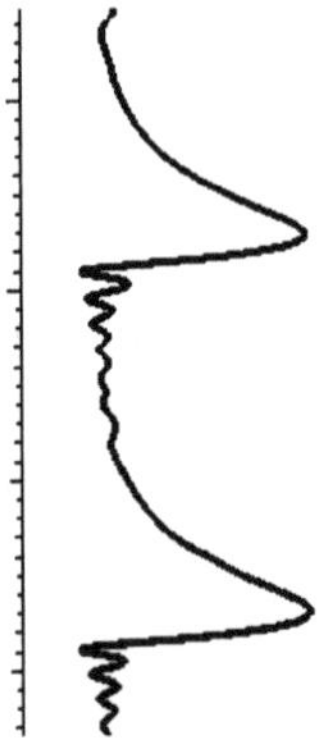

Figure 4.8: In front of the large solitary or roll waves, there are small capillary waves. (The cross-stream direction has been magnified by a factor of 50.)

Although the Marangoni force, equation (3.53), is not simulated here, it is noteworthy that it could even lead to dry patches on the heating plate. To understand this, consider a wavy film with troughs and crests. The smaller the film thickness the better is the heat transfer and the higher is the temperature at the surface. Thus the surface in the troughs is hotter than at the crests. For water, as for most liquids, the surface tension decreases with increasing temperature. The Marangoni effect tends to balance differences of the surface tension. Therefore, the water flows from the region with lower surface tension to the surface where σ has a higher value. On account of this effect, there is a flow from the troughs to the neighboring crests. Eventually, this can lead to a rupture of the falling film.

A less dramatic consequence of the Marangoni effect is the formation of rivulets. The tendency to form crests and troughs in the spanwise direction is amplified by the Marangoni force. For a picture see, e. g., Kalliadasis et al. [72], page 12.

4.6 Experimental correlations

The VDI Heat Atlas [142] is a good source to start with in order to get an overview on the heat transfer in falling films. Alekseenko et al. [13] write on many experimental aspects of falling films. Christians [32] studies heat transfer and dry-out in falling film bundle evaporators.

Wavy, laminar falling films have a smaller average thickness compared to a smooth film. To account for this fact, Kapitza [74, 75] proposed

$$\bar{\delta}_{\mathrm{lam,wav}} = \left(\frac{2.4\nu^2}{g}\right)^{1/3} \mathrm{Re}^{1/3} \tag{4.48}$$

Note that if 2.4 was replaced by 3 it would be the film thickness according to Nusselt. For turbulent falling films, Brauer [27] concluded from optical measurements:

$$\bar{\delta}_{\mathrm{turb}} = 0.302\left(\frac{3\nu^2}{g}\right)^{1/3} \mathrm{Re}^{8/15} \tag{4.49}$$

In order to understand the correlations for the heat transfer, we need to recall the definition of the Nusselt number, Nu, first. Although this has been done in Section 4.2 en passant, it will be given a bit more space here. From Fourier's law of heat conduction, equation (3.35), the heat flux q_w from the wall to the film is

$$q_w = -\lambda_L \left.\frac{\partial T}{\partial y}\right|_{y=0}. \tag{4.50}$$

Through the film, the heat is transported from the wall to the gas-liquid interface whose temperature shall be denoted by T_i. The local heat transfer coefficient α_x is defined by

$$q_w = \alpha_x (T_w - T_i). \tag{4.51}$$

The term "local" indicates that it is a function of the coordinate x. The local Nusselt number, Nu_x, is calculated from

$$\boxed{\mathrm{Nu}_x = \frac{\alpha_x}{\lambda_L}\left(\frac{\nu_L^2}{g}\right)^{1/3}}. \tag{4.52}$$

The heat transfer coefficient as well as the Nusselt number are often averaged:

$$\bar{\alpha} = \frac{1}{L}\int_0^L \alpha_x dx \qquad \text{mean heat transfer coefficient} \tag{4.53}$$

and

$$\overline{\mathrm{Nu}} = \frac{1}{L}\int_0^L \mathrm{Nu}_x dx \qquad \text{mean Nusselt number} \tag{4.54}$$

First, we consider nonevaporating falling films. For laminar, hydrodynamically and thermally developed film flow, the mean Nusselt number can be calculated from

$$\overline{\mathrm{Nu}} = C_\infty \mathrm{Re}^{-1/3}, \tag{4.55}$$

where $C_\infty = 1.3$ in the case of constant wall temperature and $C_\infty = 1.43$ for constant wall heat flux [142].

In the thermal entrance region with length L, the following relation is frequently used:

$$\overline{\mathrm{Nu}} = C_0 \sqrt[3]{\mathrm{Re}^{1/3}\mathrm{Pr}\left(\frac{\nu^2}{g}\right)^{1/3} \Big/ L} \tag{4.56}$$

The parameter C_0 is given the value 0.912 for $T_w = \text{const}$ and 1.1 for $q_w = \text{const}$.

For the transition to turbulent film flow,

$$\overline{\mathrm{Nu}} = 0.0425\mathrm{Re}^{1/5}\mathrm{Pr}^{0.344} \tag{4.57}$$

And for fully turbulent films,

$$\overline{\mathrm{Nu}} = 0.0136\mathrm{Re}^{2/5}\mathrm{Pr}^{0.344} \tag{4.58}$$

According to the VDI Heat Atlas [142], the largest value from equations (4.55) to (4.58) is employed for the determination of the mean heat transfer coefficient $\bar{\alpha}$.

Chun and Seban [33] make the following ansatz to cover wavy, laminar ("wl") as well as turbulent ("turb") film flow:

$$\mathrm{Nu} = \sqrt{\mathrm{Nu}_{\mathrm{wl}}^2 + \mathrm{Nu}_{\mathrm{turb}}^2} \tag{4.59}$$

The expressions in the square root are such that in the wavy, laminar regime $\mathrm{Nu}_{\mathrm{wl}}$ prevails and for higher Reynolds' numbers $\mathrm{Nu}_{\mathrm{turb}}$. Chun and Seban give as specification:

$$\mathrm{Nu}_{\mathrm{wl}} = 0.82(4\mathrm{Re})^{-0.22} \tag{4.60}$$

$$\mathrm{Nu}_{\mathrm{turb}} = 0.0038(4\mathrm{Re})^{0.4}\mathrm{Pr}^{0.65} \tag{4.61}$$

The correlations were obtained with water. They have proved satisfactory for Pr up to 5. In the VDI Heat Atlas [142], there are similar correlations experimentally verified for Pr up to 7.

Alhusseini et al. [14] make a different ansatz:

$$\mathrm{Nu} = \sqrt[5]{\mathrm{Nu}_{\mathrm{wl}}^5 + \mathrm{Nu}_{\mathrm{turb}}^5} \tag{4.62}$$

In the wavy, laminar regime, they also take the Kapitza number into account:

$$\mathrm{Nu}_{\mathrm{wl}} = 2.65(4\mathrm{Re})^{-0.158}\mathrm{Ka}^{0.0563} \tag{4.63}$$

They calculate $\mathrm{Nu}_{\mathrm{turb}}$ in a rather complicated manner:

$$\mathrm{Nu}_{\mathrm{turb}} = \frac{\mathrm{Pr}\delta_+^{1/3}}{a_1\mathrm{Pr}^{3/4} + a_2\mathrm{Pr}^{1/2} + a_3\mathrm{Pr}^{1/4} + b(\mathrm{Pr}\mathrm{Ka})^{1/2} + c_1 + c_2 4\mathrm{Re}} \tag{4.64}$$

where

$$a_1 = 9.17$$

$$a_2 = 1.0304\left(\frac{130 + \delta_+}{\delta_+}\right)$$

$$a_3 = 0.0289\left(\frac{152100 + 2340\delta_+ + 7\delta_+^2}{\delta_+^2}\right)$$

$$b = 2.51 \times 10^6\left(\frac{\delta_+^{0.333}}{(4\mathrm{Re})^{3.49\mathrm{Ka}^{0.0675}}}\right)\mathrm{Ka}^{-0.173}$$

$$c_1 = 8.82$$

$$c_2 = 0.0003$$

$$\delta_+ = 0.0946(4\mathrm{Re})^{0.8}$$

Alhusseini et al. used water and propylene glycol. The Prandtl number ranged from 2 to 50.

Finally, a correlation for $\mathrm{Nu}_{\mathrm{turb}}$ by *Numrich* [106] based on several measurements with different liquids covering Prandtl numbers up to 52 shall be mentioned:

$$\mathrm{Nu}_{\mathrm{turb}} = 0.003(4\mathrm{Re})^{0.44}\mathrm{Pr}^{0.4} \tag{4.65}$$

The correlation of Nosoko et al. [105] are employed to test OpenFOAM in Chapter 7. They employ laminar water films at Re = 14 ~ 90 and at temperatures of 5 ~ 23 °C. The distance between the crests of the solitary waves is determined as wavelength Δx_{wave} and the period Δt_{wave} is measured. The phase velocity is then calculated from $u_{\mathrm{ph}} = \Delta x_{\mathrm{wave}}/\Delta t_{\mathrm{wave}}$. Furthermore, the peak height, $\delta_{\max}$, is observed. Their dimensionless correlation for the phase velocity is

$$N_{\mathrm{uph}} = 1.13K_F^{0.02}N_{\Delta x\,\mathrm{wave}}^{0.31}\mathrm{Re}^{0.37} \tag{4.66}$$

where

$$N_{\mathrm{uph}} = u_{\mathrm{ph}}/(\nu g)^{1/3},$$

$$K_F = \rho^3\nu^4 g/\sigma^3 \quad \text{(physical properties group), and}$$

$$N_{\Delta x\,\mathrm{wave}} = \Delta x_{\mathrm{wave}}(g/\nu^2)^{1/3} \quad \text{(dimensionless wave distance).}$$

For the peak height, they found out that

$$N_{\delta\max} = 0.49K_F^{0.044}N_{\Delta x\,\mathrm{wave}}^{0.39}\mathrm{Re}^{0.46} \quad \text{with } N_{\delta\max} = \delta_{\max}(g/\nu^2)^{1/3} \tag{4.67}$$

It has been observed in experiments and simulation that also without excitation first two-dimensional waves appear after a certain distance from the inlet. These 2D waves become three-dimensional farther downstream.

Simulations are a good means to demonstrate recirculation in the first two troughs in front of the solitary wave. This phenomenon has been confirmed experimentally by Tihon [136].

Under certain conditions, falling films tend to break up and leave dry patches on the wall. If a wavy film is heated by the wall, the wave crest usually has a lower temperature than the wave trough. Remember, on account of the Marangoni force there is a flow from the trough to the crest which can eventually lead to a break up. Moreover, other instabilities can take place if the wall surface has regions of different wettabilities. Generally, it can be said that the phenomenon occurs if the mass flow rate is too small and/or the wall heat flux is too high. For more details see, e. g., [142, 93].

4.7 Simulations

Only some numerical experiments with falling films are mentioned here. Among other sources, the proceedings of the International Berlin Workshops on Transport Phenomena with Moving Boundaries (IBW) are cited.

Renz and his successor Kneer, together with their groups, published many papers on wavy falling films see, e. g., [5, 9, 85, 8, 7, 39]. Their work is experimental and numerical. Among other things, they found out that for a heated film the lines of constant temperature are parallel if the Prandtl number is low, otherwise they are distorted. From these observations they concluded that at low Pr the heat is mainly transported by conduction and at high Pr by convection. Dietze [38], a member of Kneer's group at that time, observed a capillary separation eddy experimentally as well as numerically. Other researchers who succeeded in simulating wavy falling films are Miyara [99, 100, 101] and Kunugi and Kino [81]. Among other things, they simulate two-dimensional waves that arise from a periodic excitement. Thus, they observe large solitary waves and, in front of the solitary waves, small capillary waves. Their simulated results are compared to the correlation of Nosoko et al. [105]; see Section 4.6 with equations (4.66) to (4.67).

Dietze et al. [40] performed three-dimensional simulations. For the direct numerical simulations, OpenFOAM was employed. Furthermore, the WRIBL model by Scheid et al. [126] was used. The results were compared to experiments. All in all, they observed complex spanwise flow structures.

The Ph. D. thesis by Doro [41] is of special interest, for he simulated evaporating falling films with OpenFOAM. His main concern was the evaporation of black liquor.

Kharangate employs the commercial software FLUENT for the computational modeling of evaporating falling films. His Ph. D. thesis is available on the Internet [76]. He, Mudawar, and Lee assume a total damping of the turbulent viscosity ν_t at the gas-liquid interface, [77].

4.8 Mixture effects

Wadekar [147] also treats mixture effects. Turbulence on account of wires can be the reason for such an effect. Figure 4.9 shows two temperature profiles inside an evaporating falling film: the upper one with mixture effect, the lower one without. Due to the evaporation of the volatile component, there is a higher concentration of the other component at the interface. In the case of seawater this means that the water evaporates and the salt is left inside the liquid film. The salt concentration at the interface is higher than inside the bulk before the evaporation. A higher concentration of solved solids results in a boiling point elevation. Thus by the mixture effect the wall superheat is reduced.

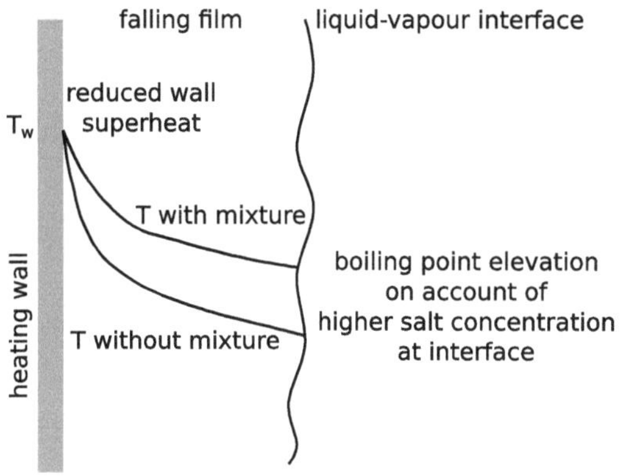

Figure 4.9: Reduction of wall superheat due to mixture effect.

4.9 Enhancement of heat transfer

To learn about the heat transfer enhancement in falling films and experiments on structured surfaces, the Ph. D. thesis by Lozano Aviles [93] is a good source. The papers by Gambaryan-Roisman and Stephan [53, 54, 55] shall also be recommended.

There are active and passive methods to enhance the heat transfer. Active ones are, e. g., mechanical that cause the wall or the liquid to vibrate. Passive methods include, e. g., structured heating surfaces or additives to the liquid that stabilize the film. The wall may be structured by longitudinal grooves (sinusoidal, triangular, or rectangular) or fins. The research of Gambaryan-Roisman and Stephan with longitudinal grooves has shown that the area where the film is extremely thin is very favorable to the heat transfer. They observed a stabilizing effect in the case of longitudinal rectangular grooves. Lozano Aviles performed experimental studies with longitudinal grooves and fins, where she found a better heat transfer. However, the improvement was difficult to specify since it depends on fluid properties. Furthermore, she observed that the longitudinal grooves prevent the film from breaking down and leaving dry patches that otherwise would spread in the horizontal direction. All in all, experimental studies are by far more numerous than theoretical ones.

4.10 Harmonic waves

A sinusoidal wave shall be called "harmonic" in the following. This terminology rather originates from a physicist's point of view. Instead of "harmonic wave" the term "linear kinematic wave" is often used in the literature on falling films, [13, 72]. Before we can formulate the wave equation, we have to define the specific flow rate per unit of the film width:

$$q_u = \int_0^\delta u\,dy = \overline{u}\delta \tag{4.68}$$

Conservation of the flow rate yields

$$\frac{\partial \delta}{\partial t} + \frac{\partial q_u}{\partial x} = 0. \tag{4.69}$$

With the aid of the chain rule, this can be written as

$$\frac{\partial \delta}{\partial t} + \frac{\partial q_u}{\partial \delta}\frac{\partial \delta}{\partial x} = 0. \tag{4.70}$$

We can interpret $\partial q_u/\partial \delta$ as the phase velocity u_{ph} of the wave. If a Nusselt velocity profile is assumed, we obtain

$$u_{\mathrm{ph}} = \frac{\partial q_u}{\partial \delta} = \frac{\partial}{\partial \delta}\left(\frac{g}{3\nu_L}\delta^3\right) = \frac{g}{\nu_L}\delta^2 = 3\overline{u}_{\mathrm{Nusselt}} \tag{4.71}$$

Thus we have the following wave equation:

$$\frac{\partial \delta}{\partial t} + u_{\mathrm{ph}}\frac{\partial \delta}{\partial x} = 0. \tag{4.72}$$

Two possibilities to modulate the film are considered here:

$$\text{ansatz 1:}\quad \delta(x,t) = \delta_{\mathrm{Nusselt}}(1 + \varepsilon\sin(\omega t - k_x x)), \tag{4.73}$$

where ε is the modulation amplitude (not to be confounded with the dissipation of the turbulent kinetic energy). Equation (4.73) as well as equation (4.74) fulfil the wave equation, equation (4.72). The subscript "Nusselt" refers to a smooth film. The angular frequency is denoted by ω, k_x is the angular wave number. In equation (4.73), the film thickness is directly modulated, whereas in equation (4.74) the flow rate is varied.

$$\begin{aligned} &\text{ansatz 2:} \\ &q_u(x,t) = q_{u,\text{Nusselt}}(1 + \varepsilon \sin(\omega t - k_x x)) \\ \Rightarrow \quad &\delta(x,t) = \left(\frac{3\nu_L q_u(x,t)}{g}\right)^{1/3} \end{aligned} \tag{4.74}$$

Both solutions yield the important relation

$$u_{\text{ph}} = \frac{\omega}{k_x} = \frac{2\pi f}{2\pi/\lambda_x} = \lambda_x f, \tag{4.75}$$

where f is the frequency and λ_x denotes the wavelength.

In Section 4.4, we encountered harmonic waves as a perturbation of a smooth film. In Chapter 7, a film is excited harmonically at the inlet. In Section 9.5, harmonic waves will be employed to refine the algorithms for a smooth film.

4.11 Long waves

For a long time, it was computationally too expensive to solve the full Navier–Stokes equations if a laminar wavy falling films was to be simulated. Therefore, engineers sought for simpler equations. Here, only the pioneering papers of Kapitza [74, 75] and Demekhin and Shkadov [36] shall be mentioned. At the end of this section, we will have two equations which are called *long wave equations* in the remainder of the book. A derivation is only roughly outlined here, for details see Alekseenko et al. [13].

A scale Λ_{lw} is introduced which is in the order of the wavelength λ_x. As we are concerned with long waves, the wavelength is much larger than the thickness δ_0 of the corresponding smooth film. Thus a small parameter can be defined:

$$\varepsilon_{\text{lw}} = \frac{\delta_0}{\Lambda_{\text{lw}}} \ll 1 \tag{4.76}$$

The starting point for the long wave equations are the Navier–Stokes equations with boundary conditions. Only two spatial dimensions are considered here although a generalization to 3D is possible. The coordinates, the time, the velocities, the pressure, and the flow rate are scaled to make the equations dimensionless. In order to solve the equations approximately, the following ansatz is made:

$$\psi = \psi_0 + \varepsilon_{\text{lw}}\psi_1 + \varepsilon_{\text{lw}}^2\psi_2 + \varepsilon_{\text{lw}}^3\psi_3 + \cdots \tag{4.77}$$

The symbol ψ denotes a quantity for which an approximate solution is to be found, e. g., the streamwise velocity u. Since ε_{lw} is very small, the series is expected to converge rapidly. Therefore, only the lower orders in the expansion are usually calculated.

The resulting long wave equations are two coupled differential equations for the film thickness δ and the specific flow rate per unit of the film width q_u. On account of the integration in the definition of q_u, equation (4.68), many researchers speak of an "integral method." The cross-stream velocity v is completely neglected. The streamwise velocity u is averaged. For a vertically falling film, the *long wave equations* are

$$\begin{aligned} \frac{\partial q_u}{\partial t} + \frac{6}{5}\frac{\partial}{\partial x}\left(\frac{q_u^2}{\delta}\right) &= -3\frac{\nu q_u}{\delta^2} + g\delta + \frac{\sigma}{\rho}\delta\frac{\partial^3\delta}{\partial x^3} \\ \frac{\partial \delta}{\partial t} &= -\frac{\partial q_u}{\partial x} \end{aligned} \tag{4.78}$$

This is equivalent to

$$\begin{aligned} \frac{\partial q_u}{\partial t} &= -\frac{12}{5}\frac{q_u}{\delta}\frac{\partial q_u}{\partial x} + \frac{6}{5}\frac{q_u^2}{\delta^2}\frac{\partial \delta}{\partial x} - 3\frac{\nu q_u}{\delta^2} + g\delta + \frac{\sigma}{\rho}\delta\frac{\partial^3\delta}{\partial x^3} \\ \frac{\partial \delta}{\partial t} &= -\frac{\partial q_u}{\partial x} \end{aligned} \tag{4.79}$$

Note that the long wave equations are presented dimensionally. One assumption to equation (4.79) is a self-similar, semiparabolic velocity profile, like equation (4.82). The Reynolds' number should be in the range

$$1 \lesssim \mathrm{Re} \lesssim \frac{1}{\varepsilon_{\mathrm{lw}}^2}. \tag{4.80}$$

In Section 9.4, the long wave equations will be discretized and numerical solutions for q_u and δ calculated.

4.12 Zero streamline

There is a special streamline in the flow behind an obstacle that is attached to a wall. This zero streamline is depicted in Figure 4.10 for a wire with rectangular cross-section. Through the zero streamline, there is no heat and mass transport at a macroscopic scale. It ends with the stagnation point, where the wall shear stress is zero,

$$\tau_w = \mu_L \left.\frac{\partial u}{\partial y}\right|_{y=0} = 0 \quad \Rightarrow \quad \left.\frac{\partial u}{\partial y}\right|_{y=0} = 0 \tag{4.81}$$

The velocity component u changes the sign in the separation region next to the wall.

4.13 Reasonable approximations

What do we take from this chapter in order to write our own programs, like this is done in Chapter 9? If we cannot simulate an evaporating saltwater falling film on a large scale, are reasonable approximations possible? The answer is yes.

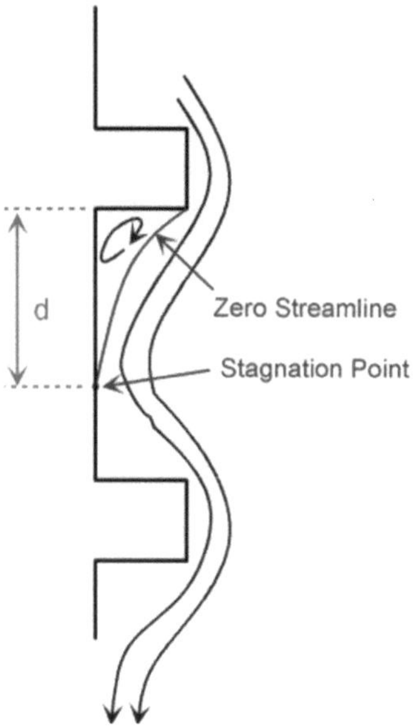

Figure 4.10: The zero streamline behind a turbulence wire with rectangular cross-section.

– As the streamwise velocity component dominates, the velocity profile in a falling film is approximately unidirectional and semiparabolic.

$$u(x,y) = 2u_s(x)\left[\frac{y}{\delta(x)} - \frac{1}{2}\left(\frac{y}{\delta(x)}\right)^2\right] \tag{4.82}$$

The velocity at the surface, $u_s(x)$, can be obtained from the specific flow rate q_u, equation (4.68). Integration of equation (4.82) yields

$$\overline{u}(x) = \frac{1}{\delta}\int_0^{\delta} u(x,y)\,dy = \frac{2}{3}u_s(x) \tag{4.83}$$

From the average velocity and the definition of the specific flow rate, the surface velocity follows:

$$u_s(x) = \frac{3}{2}\frac{q_u(x)}{\delta(x)}. \tag{4.84}$$

If $q_u(x)$ is not available, the Nusselt solution for a smooth film can be of help:

$$u_s(x) = \frac{1}{2}\frac{\rho_L - \rho_G}{\mu_L}g(\delta(x))^2 \tag{4.85}$$

Usually the density of the gas is negligible. So equation (4.85) yields

$$u_s(x) = \frac{g}{2\nu_L}(\delta(x))^2. \tag{4.86}$$

According to Dietze, [38] Figures 2.12 and 2.13, equation (4.82) describes well the numerical data. The surface velocity is taken from the direct numerical simulation. The Nusselt solution for a smooth film, on the other hand, is not so accurate.
- The temperature profile in an evaporating film is close to be linear, starting with the wall temperature and having the saturation temperature at the surface.
- The entrance region is expected to be rather short.

5 Numerical methods

In this chapter, I shall rather concentrate on the numerical methods actually used than give an overview of all methods that exist. Numerics is a field that reaches much further than applied mathematics. With the development of more and more powerful computers, it has established as an important tool of science beside of experiment and abstract theory.

5.1 Finite volumes

The finite volume method is one way to solve differential equations, ordinary as well as partial. It is employed in the CFD softwares Star-CD and OpenFOAM and shall therefore be explained in detail a little more. The other two frequent methods, finite elements and finite differences shall be briefly outlined afterwards. For a detailed introduction to the finite volume method, see Versteeg and Malalasekera [143, 144]. A book that is supplemented by finite volume source code is Ferziger and Peric [50, 1]. The finite difference method is thoroughly explained in Anderson [17]. Chung [34] treats finite differences as well as finite elements. Zienkiewicz et al. [156, 155] treat the finite element method in-depth in several volumes.

Let us first state a typical transport equation in differential form:

$$\frac{\partial(\rho\phi)}{\partial t} + \frac{\partial(\rho\phi u)}{\partial x} + \frac{\partial(\rho\phi v)}{\partial y} + \frac{\partial(\rho\phi w)}{\partial z} = \operatorname{div}(\Gamma_\phi \operatorname{grad} \phi) + R_\phi. \tag{5.1}$$

If this model equation stands for the continuity equation, $\phi = 1$ and $\Gamma_\phi = R_\phi = 0$. To represent the x-momentum equation $\phi = u$, $\Gamma_\phi = \mu$ (dynamic viscosity) and $R_\phi = -\partial p/\partial x + \rho f_x$, analogous expressions for the y and z momentum equations, cf. Equations (3.22) to (3.24). The above transport equation can also be regarded as the simplified energy equation, equation (3.30). In this case $\phi = c_V T$, $\Gamma_\phi = \lambda/c_V$, and $R_\phi = 0$.

In the finite volume method the model equation is integrated over a control volume V which is usually a cell of the numerical grid. Thus equation (5.1), with the aid of Gauss' theorem, becomes

$$\int_V \frac{\partial(\rho\phi)}{\partial t}\, dV + \int_A \rho\phi \boldsymbol{U}\, \boldsymbol{dA} + = \int_A \Gamma_\phi \operatorname{grad} \phi\, \boldsymbol{dA} + \int_V R_\phi\, dV. \tag{5.2}$$

In equation (5.2), the control surface is denoted by A. The equation above contains derivatives and integrals. To see how these mathematical entities are approximated in numerical analysis, consider the function $f(x)$. If the derivative is to be evaluated, the difference quotient is employed:

$$\frac{df}{dx} = \lim_{\Delta x \to 0} \frac{\Delta f}{\Delta x} \approx \frac{\Delta f}{\Delta x} \quad \text{for small } \Delta x \tag{5.3}$$

https://doi.org/10.1515/9783110592337-005

If an integral is to be approximated, it is important to have a small control volume. In such a case, the kernel can be regarded as constant.

$$\int_{\Delta x} f\,dx \approx f(x_0)\int_{\Delta x} dx = f(x_0)\cdot\Delta x \quad \text{with } x_0 \in \Delta x \text{ and } \Delta x \text{ small} \tag{5.4}$$

5.1.1 Diffusion

In order to see how the finite volume method finds an approximate solution of a differential equation, let us consider the stationary one-dimensional diffusion equation:

$$\frac{d}{dx}\left(\Gamma\frac{d\phi}{dx}\right) + R = 0 \tag{5.5}$$

The first step is to generate a numerical grid. Since the problem is one-dimensional, there is a line to be divided into segments, the "cells." Figure 5.1 shows one cell. Its center is denoted by the letter "P." The center of its neighbor to the left is marked with the letter "W" like "West." In an analogous manner, the neighbor to the right is "E" for "East." The left cell face is referred to as "w" and the right one as "e."

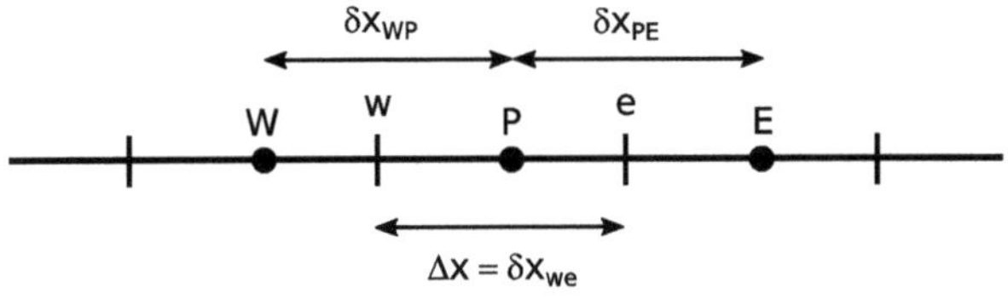

Figure 5.1: One-dimensional control volume inside a numerical grid.

The main step of the finite volume method is the integration over a control volume ΔV. This control volume shall be the cell with center P. Thus $\Delta V = \Delta x \cdot A$ (cf. Figure 5.1),

$$\int_{\Delta V}\frac{d}{dx}\left(\Gamma\frac{d\phi}{dx}\right)dV + \int_{\Delta V} R\,dV = \left(\Gamma A\frac{d\phi}{dx}\right)_e - \left(\Gamma A\frac{d\phi}{dx}\right)_w + \bar{R}\Delta V = 0 \tag{5.6}$$

It is seen from equation (5.6) that the coefficient Γ and the derivative $d\phi/dx$ are to be evaluated at the cell faces. For this reason, we assume that the values are known at the cell centers P, W, and E. For Γ a linear approximation is chosen,

$$\Gamma_w = \frac{\Gamma_W + \Gamma_P}{2} \quad \text{und} \quad \Gamma_e = \frac{\Gamma_P + \Gamma_E}{2} \tag{5.7}$$

For the derivative $d\phi/dx$, a scheme shall be employed that is called "central differences" or "central discretization" (CD). Thus the diffusive currents become

$$\left(\Gamma A\frac{d\phi}{dx}\right)_e = \Gamma_e A_e\left(\frac{\phi_E - \phi_P}{\delta x_{\mathrm{PE}}}\right) \tag{5.8}$$

$$\left(\Gamma A\frac{d\phi}{dx}\right)_w = \Gamma_w A_w\left(\frac{\phi_P - \phi_W}{\delta x_{\mathrm{WP}}}\right) \tag{5.9}$$

The source term in equation (5.6) is modeled by the ansatz

$$\bar{R}\Delta V = R_u + R_P\phi_P \tag{5.10}$$

By substituting the expressions (5.7), (5.8), (5.9), and (5.10) into equation (5.6), we obtain

$$\Gamma_e A_e\left(\frac{\phi_E - \phi_P}{\delta x_{\mathrm{PE}}}\right) - \Gamma_w A_w\left(\frac{\phi_P - \phi_W}{\delta x_{\mathrm{WP}}}\right) + (R_u + R_P\phi_P) = 0 \tag{5.11}$$

The terms in equation (5.11) can be placed in a different order:

$$\left(\frac{\Gamma_e}{\delta x_{\mathrm{PE}}}A_e + \frac{\Gamma_w}{\delta x_{\mathrm{WP}}}A_w - R_P\right)\phi_P = \left(\frac{\Gamma_w}{\delta x_{\mathrm{WP}}}A_w\right)\phi_W + \left(\frac{\Gamma_e}{\delta x_{\mathrm{PE}}}A_e\right)\phi_E + R_u \tag{5.12}$$

In an abbreviated manner, we write

$$a_P\phi_P = a_W\phi_W + a_E\phi_E + R_u \tag{5.13}$$

with

$$a_W = \frac{\Gamma_w}{\delta x_{\mathrm{WP}}}A_w; \quad a_E = \frac{\Gamma_e}{\delta x_{\mathrm{PE}}}A_e; \quad a_P = a_W + a_E - R_P.$$

A discretization equation has been found for a cell P. If the coefficients and sources are known for each cell of the numerical grid, usually after applying boundary conditions, a (sparse) system of linear equations is obtained that can be solved by means of applied mathematics. In our case, the Thomas algorithm would be very efficient, which essentially is the Gauss algorithm with diagonal strategy; cf. [17]. For a general account of numerical mathematics see, e. g., Köckler and Schwarz [82] (in German). If a reader is rather interested in "numerical recipes," Press et al. [113] is recommended. In this book, the algorithms are explained in a less formal language. A book solely dedicated to direct and iterative solutions of systems of linear equations is Kanzow [73] (in German). For iterative methods for large sparse systems, see Hackbusch [60]. The iterative algorithm of Stone [134] is often employed in the programs of Peric [1]. For this monograph, Stone's algorithm was used for two-dimensional problems.

5.1.2 Convection

To proceed further, we treat a one-dimensional stationary convection-diffusion equation:

$$\frac{d}{dx}(\rho u\phi) = \frac{d}{dx}\left(\Gamma\frac{d\phi}{dx}\right) \tag{5.14}$$

Integration over a cell volume yields

$$(\rho uA\phi)_e - (\rho uA\phi)_w = \left(\Gamma A\frac{d\phi}{dx}\right)_e - \left(\Gamma A\frac{d\phi}{dx}\right)_w \tag{5.15}$$

Now we have to evaluate ϕ_e and ϕ_w, i. e., the values of the scalar ϕ at the eastern and western face of the cell. In the case of a uniform grid central discretization (CD) would calculate the face values as

$$\phi_e = (\phi_P + \phi_E)/2 \quad \text{and} \quad \phi_w = (\phi_W + \phi_P)/2 \tag{5.16}$$

Apart from numerical instabilities that may occur CD is not always the discretization scheme of choice for convective fluxes. Apparently, the flow direction should be taken into account, i. e., the upstream field should have a stronger influence than the field downstream. The "upwind discretization scheme" (UD) accomplishes this physically motivated property. In our case, it is defined by

$$\phi_w = \phi_W \quad \text{and} \quad \phi_e = \phi_P \quad \text{for } u_w > 0 \text{ and } u_e > 0 \tag{5.17}$$

If the flow has the other direction,

$$\phi_w = \phi_P \quad \text{and} \quad \phi_e = \phi_E \quad \text{for } u_w < 0 \text{ and } u_e < 0 \tag{5.18}$$

UD is numerically more stable than CD giving some solution also when the mesh is coarse. This is the reason why UD is the default discretization scheme in most CFD programs. However, UD suffers from *false diffusion* if the flow is oblique with respect to the grid lines. In this case, a refinement of the mesh helps to make the solution more accurate.

5.1.3 Transient problems

Finally, the possibilities shall be explained that we have to solve transient problems. For this reason, we shall treat the time-dependent one-dimensional heat conduction equation:

$$\rho c\frac{\partial T}{\partial t} = \frac{\partial}{\partial x}\left(\lambda\frac{\partial T}{\partial x}\right) + R \tag{5.19}$$

In equation (5.19), c denotes the specific heat. (In the case of a solid or a liquid $c_V = c_p = c$.) Equation (5.19) has been solved numerically in the one-dimensional (1D) simulations of which the results are presented in Chapter 9. For the finite volume method, equation (5.19) has to be integrated over a cell volume and a time interval:

$$\int_t^{t+\Delta t}\int_V \rho c\frac{\partial T}{\partial t}\,dV\,dt = \int_t^{t+\Delta t}\int_V \frac{\partial}{\partial x}\left(\lambda\frac{\partial T}{\partial x}\right)dV\,dt + \int_t^{t+\Delta t}\int_V R\,dV\,dt \tag{5.20}$$

Since the problem is 1D, equation (5.20) can be written as

$$\int_w^e \left[\int_t^{t+\Delta t} \rho c \frac{\partial T}{\partial t} dt \right] dV = \int_t^{t+\Delta t} \left[\left(\lambda A \frac{\partial T}{\partial x} \right)_e - \left(\lambda A \frac{\partial T}{\partial x} \right)_w \right] dt + \int_t^{t+\Delta t} \bar{R} \Delta V \, dt \tag{5.21}$$

Under the assumption that the cell is small, the temperature at node P prevails in the whole cell,

$$\int_w^e \left[\int_t^{t+\Delta t} \rho c \frac{\partial T}{\partial t} dt \right] dV = \rho c (T_P - T_P^0) \Delta V \tag{5.22}$$

In equation (5.22), the upper index "0" refers to the time t. The time $t+\Delta t$ has no index. The same result as in equation (5.22) would be obtained if we substituted $\partial T/\partial t$ with $(T_P - T_P^0)/\Delta t$. Employing the central discretization scheme for the spatial derivatives, we get

$$\begin{aligned} \rho c (T_P - T_P^0) \Delta V = & \int_t^{t+\Delta t} \left[\left(\lambda_e A \frac{T_E - T_P}{\delta x_{\mathrm{PE}}} \right) - \left(\lambda_w A \frac{T_P - T_W}{\delta x_{\mathrm{WP}}} \right) \right] dt \\ & + \int_t^{t+\Delta t} \bar{R} \Delta V \, dt \end{aligned} \tag{5.23}$$

The question remains how to evaluate the temperature during the time interval Δt which is assumed to be very small. We could take the "old" temperature T_P^0, the "new" temperature T_P, or a combination of both. We therefore introduce the parameter $\theta \in [0,1]$:

$$I_T = \int_t^{t+\Delta t} T_P \, dt = [\theta T_P + (1-\theta) T_P^0] \Delta t \tag{5.24}$$

We consider three cases:

$$I_T = \begin{cases} T_P^0 \Delta t: & \theta = 0 \\ \frac{1}{2}(T_P + T_P^0)\Delta t: & \theta = \frac{1}{2} \\ T_P \Delta t: & \theta = 1 \end{cases}$$

The scheme with $\theta = 0$ is called *explicit*. If θ is greater than zero, we speak of an *implicit* scheme. The case $\theta = 1/2$ bears a special name: it is called "Crank–Nicolson scheme." If $\theta = 1$, the scheme is referred to as *fully implicit*.

With the aid of the parameter θ, we can write equation (5.23) as

$$\begin{aligned} a_P T_P = & a_W [\theta T_W + (1-\theta) T_W^0] + a_E [\theta T_E + (1-\theta) T_E^0] \\ & + [a_P^0 - (1-\theta) a_W - (1-\theta) a_E] T_P^0 + b \end{aligned} \tag{5.25}$$

where

$$a_P = \theta(a_W + a_E) + a_P^0$$

and

$$a_P^0 = \rho c \frac{\Delta x}{\Delta t}$$

and

$$a_W = \lambda_w / \delta x_{\mathrm{WP}}; \quad a_E = \lambda_e / \delta x_{\mathrm{PE}}; \quad b = \bar{R}\Delta x.$$

The explicit scheme is conditionally stable, i. e., the time step Δt must not exceed a certain size. In our case,

$$\Delta t < \rho c \frac{(\Delta x)^2}{2\lambda} \tag{5.26}$$

The above condition has been obtained by requiring that all coefficients in equation (5.25) should be positive. For a thorough discussion on this matter, see Anderson [17]. Beside this disadvantage, the explicit scheme is easy to program.

The Crank–Nicolson scheme is more difficult to be implemented into a program. Furthermore, it is also conditionally stable for

$$\Delta t < \rho c \frac{(\Delta x)^2}{\lambda} \tag{5.27}$$

In comparison to the explicit scheme, a factor 2 is gained.

At last, the fully implicit scheme is unconditionally stable although some effort has to be done to program it. On account of its stability it is the default temporal discretization scheme in multi-purpose CFD-codes like Star-CD and OpenFOAM.

In connection with the numerical solution of transient flows, the Courant number Co is very important. Special care must be taken that during one time step the flow does not proceed a spatial distance larger than the cell size. In 1D, the Courant number may be written as

$$\boxed{\mathrm{Co} = \frac{|u| \cdot \Delta t}{\Delta x} \overset{!}{<} 1} \tag{5.28}$$

For a definition that is valid on 3D unstructured meshes see, e. g., the Ph. D. thesis by Rusche [125]. In order to avoid mathematical errors, the Courant number should be kept smaller than one.

5.2 The finite difference method

The finite difference method is even more straightforward than the finite volume method. For finite differences, the differential equations are discretized, not the integral equations. Apart from this, finite differences and finite volumes are very similar. The spatial discretization schemes CD and UD also exist with finite differences. For the temporal discretization, there is also the distinction between explicit and implicit schemes.

The discretization schemes CD, UD, and temporal explicit shall be detailed here, since they are employed in the original programs, Chapter 9. Let us consider a point "P" in a 1D grid; see Figure 5.1. Then the central discretization (CD) of the spatial derivative of a scalar ϕ is

$$\left(\frac{d\phi}{dx}\right)_P^{\text{CD}} = \frac{\phi_E - \phi_W}{2\Delta x}. \tag{5.29}$$

If the upwind discretization (UD) is employed, a backward differencing is used for a velocity greater than zero:

$$\left(\frac{d\phi}{dx}\right)_P^{\text{UD}} = \frac{\phi_P - \phi_W}{\Delta x} \quad \text{for } u_P > 0 \tag{5.30}$$

If the flow has the other direction, forward differencing is employed:

$$\left(\frac{d\phi}{dx}\right)_P^{\text{UD}} = \frac{\phi_E - \phi_P}{\Delta x} \quad \text{for } u_P < 0 \tag{5.31}$$

In order to see how a temporal discretization is done in the finite difference method, let us consider the heat conduction equation in 1D with constant material properties:

$$\frac{\partial T}{\partial t} = D_{\text{th}} \frac{\partial^2 T}{\partial x^2} \tag{5.32}$$

In this equation, T denotes the temperature and D_{th} is the thermal diffusivity. For the spatial derivative CD shall be used,

$$\left(\frac{\partial^2 T}{\partial x^2}\right)_P = \frac{T_E - 2T_P + T_W}{(\Delta x)^2} \tag{5.33}$$

The temporal derivative is approximated by the difference quotient:

$$\left(\frac{\partial T}{\partial t}\right)_P = \frac{T_P - T_P^0}{\Delta t} \tag{5.34}$$

The superscript "0" refers to the previous time. Now, the explicit temporal discretization of equation (5.32) reads

$$\frac{T_P - T_P^0}{\Delta t} = D_{\text{th}} \frac{T_E^0 - 2T_P^0 + T_W^0}{(\Delta x)^2}. \tag{5.35}$$

Note that in the spatial derivative the temperatures of the previous time step are taken. Equation (5.35) can be easily solved for the temperature T at point "P" for the updated time

$$T_P = T_P^0 + \Delta t \cdot D_{\text{th}} \cdot \frac{T_E^0 - 2T_P^0 + T_W^0}{(\Delta x)^2}. \tag{5.36}$$

For the finite difference method, the grid has to be Cartesian or, at least, transformable to a Cartesian grid. That is the reason why rather complicated geometries cannot be treated with it.

5.3 The finite element method

The finite element method requires a higher degree of mathematical sophistication than finite differences and finite volumes. With the finite volume method, it has in common the ability to be applied to unstructured meshes. Therefore, it is mostly employed for cases with complicated geometries. It shall not be further treated here. Beside of the literature cited above, the very pedagogical approach of Evans et al. [47] is recommended and the book by Zienkiewicz and Morgan [154] mentioned for its bargain price.

5.4 Volume of fluid (VOF)

There are various methods to treat a moving interface. In surface methods, the position of the interface can be marked with massless particles. In this case, a redistribution of the marker particles has to take place in general. Alternatively, the interface can be attached to the surface of some cells. Here, the topology of the mesh has to be changed according to the movement of the interface. In literature, the surface methods are classified as "surface fitting" in contrast to "surface capturing" volume methods [139]. A special volume method is the volume of fluid (VOF) method by Hirt and Nichols [65]. A scalar, c_{VOF}, is introduced that gives the volume fraction of the heavier fluid. For the sake of convenience, we shall regard the heavier fluid as liquid (L) and the lighter one as gas (G). For a cell of the numerical grid, we have

$$c_{\text{VOF}} = \frac{V_{L,\text{cell}}}{V_{\text{cell}}} \tag{5.37}$$

Thus, for a cell completely filled with liquid $c_{\text{VOF}} = 1$. If there is only gas, $c_{\text{VOF}} = 0$. The interface is located inside cells with $0 < c_{\text{VOF}} < 1$. Figure 5.2 gives an example. In cells with interface, the fluid properties are interpolated according to the value of c_{VOF}. For example, the dynamic viscosity, μ, is calculated from

$$\mu = c_{\text{VOF}} \cdot \mu_L + (1 - c_{\text{VOF}}) \cdot \mu_G. \tag{5.38}$$

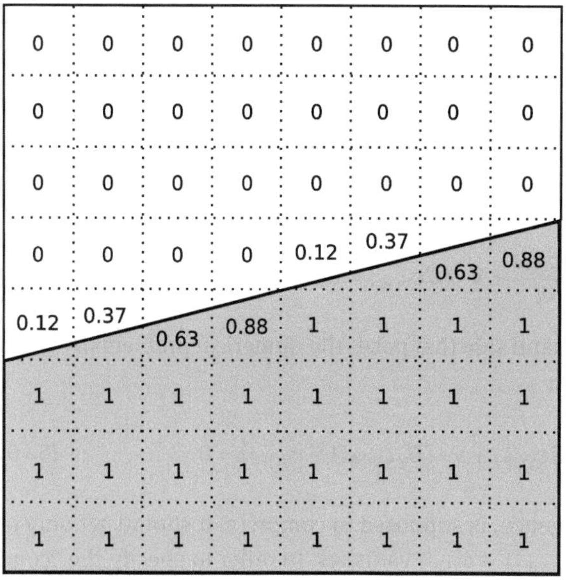

Figure 5.2: Example of the distribution of c_{VOF}.

The volume of fluid method has the advantage of being conceptually easy and robust. So it is no problem to treat merging and rupturing interfaces. However, from a numerical point of view it is a challenging task to keep the interface sharp. Ubbink [139] discusses various ways to advect and reconstruct an interface. Noh and Woodward [104] employed the Simple Line Interface Calculations (SLIC) method. In this method, the interface in a cell is approximated by a straight line. This line is either horizontal or vertical depending on the direction of the sweep. For a sweep in the x direction only, the c_{VOF}-values of the left and right neighbor cells are taken into account. For a y-sweep, the upper and lower cells are used. Chorin [31] considered the c_{VOF}-values of all four neighbor cells for each sweep. Lötstedt [94] and Youngs [152] have oblique lines for each interface cell. (For a detailed description of the Youngs algorithm, see the Appendix of the paper by Rudman [124].) For an account of methods to reconstruct the interface normal $\boldsymbol{n}$; read, e. g., Rider and Kothe [122]. Ashgriz and Poo [19] with the Flux Line-segment Model for Advection and Interface Reconstruction (FLAIR) constructed the line-segment at the cell faces, thus there might be a kink inside an interface cell.

The VOF method has the problem that the interface disperses due to numerical diffusion. Schemes that counteract this effect are called "compressive." Ubbink's Compressive Interface Capturing Scheme on Arbitrary Meshes (CICSAM) [139] is classified as "compressive-differencing." In principle, it is a blending between upwinded and downwinded schemes. For details, we refer to Ubbink's thesis [139] available on the Internet. CICSAM was presumably the scheme of choice for Star-CD. However, there are problems at stagnation points and in slow-flow regions.

In the remainder of this section, we shall first explain the idea of Weller's concept of "countergradient" transport [150], then we shall see an example of numerical diffusion if no provision is made [38]. The countergradient concept is an alternative to compressive-differencing and has been implemented into OpenFOAM.

For incompressible two-phase flow, the transport of the interface is guided by the equation

$$\frac{\partial c_{\mathrm{VOF}}}{\partial t} + \nabla \cdot (\boldsymbol{U} c_{\mathrm{VOF}}) = 0. \tag{5.39}$$

It is the second term on the left-hand side that poses the numerical problems. Now let us consider the following ansatz:

$$\frac{\partial c_{\mathrm{VOF}}}{\partial t} + \nabla \cdot (\boldsymbol{U} c_{\mathrm{VOF}}) + \nabla \cdot (\boldsymbol{U}_c c_{\mathrm{VOF}}(1 - c_{\mathrm{VOF}})) = 0 \tag{5.40}$$

The new term, the second divergence, is supposed to compress. It should act only in cells with interface, otherwise $c_{\mathrm{VOF}}(1 - c_{\mathrm{VOF}})$ vanishes. In order to specify the "compression velocity $\boldsymbol{U}_c$," we agree that it shall have the direction of the surface normal

$$\hat{\boldsymbol{n}} = \nabla c_{\mathrm{VOF}} / |\nabla c_{\mathrm{VOF}}|. \tag{5.41}$$

(Recall that the gradient has the direction of the steepest rise. Thus, $\hat{\boldsymbol{n}}$ points into the liquid.) Furthermore, we note that the velocity of the interface dispersion cannot be worse than the flow velocity $\boldsymbol{U}$. Therefore,

$$\boldsymbol{U}_c = c_c |\boldsymbol{U}| \frac{\nabla c_{\mathrm{VOF}}}{|\nabla c_{\mathrm{VOF}}|}, \tag{5.42}$$

where the compression coefficient c_c should be of the order one. Equation (5.42) may lead to a vanishing compression velocity if the velocity of the fluid becomes very slow. A better alternative could be

$$\boldsymbol{U}_c = c_c \max(|\boldsymbol{U}|) \frac{\nabla c_{\mathrm{VOF}}}{|\nabla c_{\mathrm{VOF}}|}, \tag{5.43}$$

where the maximum velocity over the whole computational domain is taken. However, it would be less extreme to introduce a residual velocity U_r which would prevent $\boldsymbol{U}_c$ to vanish:

$$\boldsymbol{U}_c = c_c (|\boldsymbol{U}| + U_r) \frac{\nabla c_{\mathrm{VOF}}}{|\nabla c_{\mathrm{VOF}}|}, \tag{5.44}$$

Another elegant alternative would be

$$\boldsymbol{U}_c = \min(c_c |\boldsymbol{U}|, \max(|\boldsymbol{U}|)) \frac{\nabla c_{\mathrm{VOF}}}{|\nabla c_{\mathrm{VOF}}|}, \tag{5.45}$$

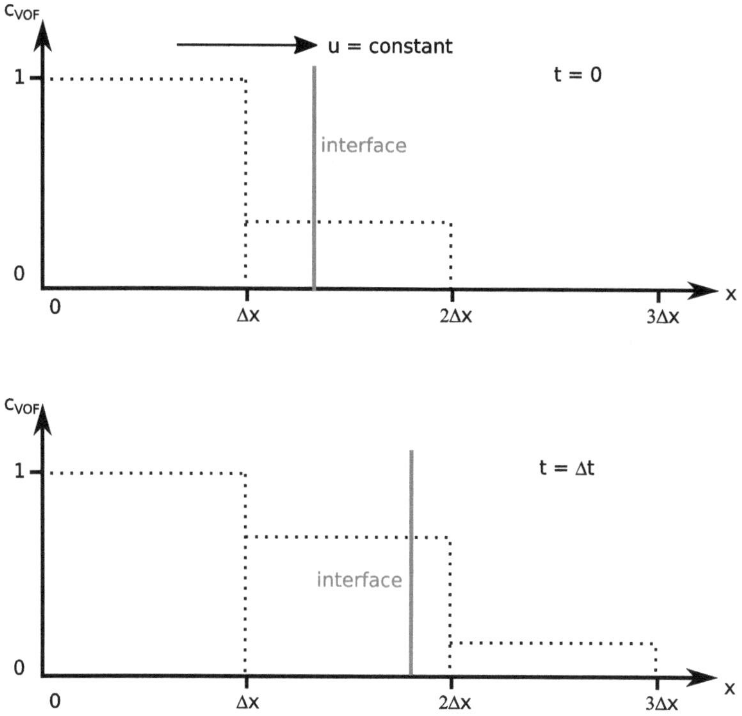

Figure 5.3: One-dimensional example of numerical diffusion in VOF.

which yields equation (5.42) in the case of $c_c \leq 1$ and approaches equation (5.43) for $c_c \to \infty$. Weller recommends $1 \leq c_c \leq 4$ although in some cases $c_c > 4$ might be useful [150].

In order to see why the VOF method tends to numerical diffusion, let us examine a simple 1D example which is depicted in Figure 5.3. In the beginning, the first cell is completely filled with liquid, the second cell is partly filled, and the third cell has no liquid at all. The interface is located at $1.3\Delta x$ where Δx is the uniform cell size. The velocity u is constant and positive. The transport equation for c_{VOF} is

$$\frac{\partial c_{\mathrm{VOF}}}{\partial t} + u\frac{\partial c_{\mathrm{VOF}}}{\partial x} = 0. \tag{5.46}$$

A possible finite difference discretization is

$$\frac{c_{\mathrm{VOF},P} - c^0_{\mathrm{VOF},P}}{\Delta t} + u\frac{c^0_{\mathrm{VOF},P} - c^0_{\mathrm{VOF},W}}{\Delta x} = 0. \tag{5.47}$$

The upwind discretization scheme and the explicit time discretization scheme are employed. The equation is solved for the new value of c_{VOF} at point P after the time step Δt:

$$c_{\mathrm{VOF},P} = c^0_{\mathrm{VOF},P} - u\Delta t\frac{c^0_{\mathrm{VOF},P} - c^0_{\mathrm{VOF},W}}{\Delta x} \tag{5.48}$$

As the Courant number Co, we choose

$$\mathrm{Co} = \frac{u\Delta t}{\Delta x} = 0.5. \tag{5.49}$$

We see from Figure 5.3 that the interface is smeared. Instead of having $c_{\mathrm{VOF}} = 0.8$ in the second cell, its value there is only 0.65. The rest of the liquid is already in the third cell ($c_{\mathrm{VOF}} = 0.15$).

5.5 Continuum surface force

As mentioned in the section on the two phase flow in Chapter 3, only the Laplace force is implemented in most multi-purpose CFD programs. Strictly speaking, the interface has to be reconstructed at each time step and the Laplace force is to be applied as some kind of boundary condition. The Continuum Surface Force (CSF) model by Brackbill et al. [25] circumvents the interface reconstruction. They treat the Laplace force as a volume force, $\boldsymbol{F}_{\mathrm{CSF}}$, in a layer around the interface. In connection with the VOF model, the iso-surface with $c_{\mathrm{VOF}} = 0.5$ is regarded as the position of the interface. The layer for the volume force $\boldsymbol{F}_{\mathrm{CSF}}$ is the region where the gradient ∇c_{VOF} does not vanish. Let us recall the Laplace force:

$$\boldsymbol{F}_{\mathrm{Laplace}} = -2\sigma H \boldsymbol{n} = -\sigma\kappa\boldsymbol{n} \tag{5.50}$$

In equation (5.50), we have substituted the mean curvature by the curvature $\kappa = 2H$. Note that $\boldsymbol{F}_{\mathrm{Laplace}}$ has the unit of a force per area. The quantity $\sigma\kappa$ may be better known as Laplace pressure. From the gradient ∇c_{VOF}, the unit vector $\boldsymbol{n}$, normal to the interface pointing from the liquid into the gas, can be approximated.

$$\boldsymbol{n} = -\frac{\nabla c_{\mathrm{VOF}}}{|\nabla c_{\mathrm{VOF}}|} \tag{5.51}$$

The curvature κ is evaluated from the divergence of $\boldsymbol{n}$:

$$\kappa = \nabla \cdot \boldsymbol{n} = -\nabla \cdot \left(\frac{\nabla c_{\mathrm{VOF}}}{|\nabla c_{\mathrm{VOF}}|} \right) \tag{5.52}$$

(The proof, why in this case ∇ can be used instead of ∇_s, can be found in the Appendix of [25].) Figure 5.4 illustrates the gradient ∇c_{VOF} and the curvature κ. Equation (5.52) is not the unique method to determine the curvature. See, e. g., the paper by Bullard et al. for an alternative [28].

Now, the expression for $\boldsymbol{F}_{\mathrm{CSF}}$, with the unit of a force per volume, in the framework of the VOF model is

$$\boldsymbol{F}_{\mathrm{CSF}} = -\sigma\left(\nabla \cdot \left(\frac{\nabla c_{\mathrm{VOF}}}{|\nabla c_{\mathrm{VOF}}|} \right)\right)(\nabla c_{\mathrm{VOF}}) \tag{5.53}$$

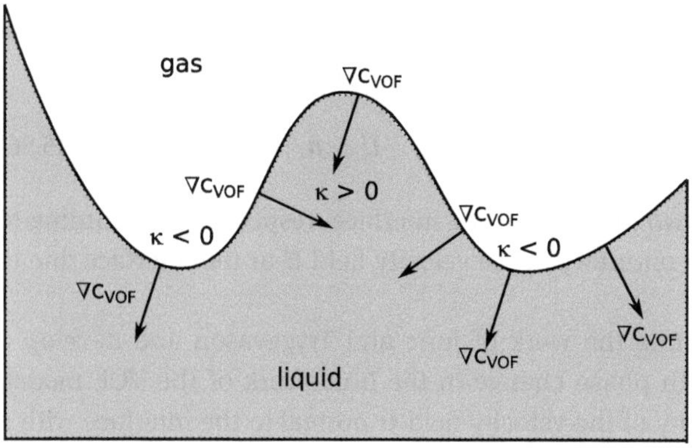

Figure 5.4: The direction of the gradient ∇c_{VOF} and the sign of the curvature κ.

Thus, by the knowledge of ∇c_{VOF}, without interface reconstruction, it is possible to simulate the Laplace force. (The reader may wonder if a factor $|\nabla c_{\text{VOF}}|^{-1}$ is missing in equation (5.53). A very simple answer can be that with this factor, the unit of $\boldsymbol{F}_{\text{CSF}}$ would be that of a force per area and that the reconstruction of the interface would be required.) In order to see how the volume force $\boldsymbol{F}_{\text{CSF}}$ is built into the Navier–Stokes equations, they are summarized in vector notation:

$$\rho \frac{\partial \boldsymbol{U}}{\partial t} + \rho(\boldsymbol{U} \cdot \nabla)\boldsymbol{U} + \nabla p = \mu \nabla^2 \boldsymbol{U} + \rho \boldsymbol{g} + \boldsymbol{F}_{\text{CSF}} \tag{5.54}$$

The CSF model in connection with VOF has the disadvantage that *parasitic currents* can occur. In order to understand how this numerical error could lead to erroneous simulations, imagine a spherical droplet at rest in a stagnant lighter fluid inside a control volume where there is no gravitational field. It is clear that in this case the droplet should not move forever. The surface tension force is exactly balanced by the pressure gradient. However, in a VOF simulation small perturbations are initiated by tiny numerical inaccuracies. The droplet starts to wobble and to move. Harvie et al. analyzed this purely numerical effect [63]. They conclude with a single correlation for the maximum parasitic current magnitude:

$$U_{\text{parasitic}} = \min(c_t U_t, c_a U_a, c_v U_v), \tag{5.55}$$

where U_t, U_a, U_v are the transient, advective and viscous imbalance, respectively (they are evaluated in [63]). The parameters c_t, c_a, and c_v depend on the specific implementation of the CSF model.

5.6 Flows with phase change

The papers by Juric and Tryggvason [69] and by Welch and Wilson [149] have been some sort of inspiration for the numerical modeling of liquid-gas phase changes. Juric and Tryggvason use a Lagrange method with marker particles for the localization of the interface. The following relation holds for the mass flux at the interface:

$$\dot{m} = \rho_l(\boldsymbol{U}_l - \boldsymbol{U}_i) \cdot \boldsymbol{n} = \rho_g(\boldsymbol{U}_g - \boldsymbol{U}_i) \cdot \boldsymbol{n}, \tag{5.56}$$

where l, g and i refer to the two phases and the interface, respectively. According to equation (5.56) there is a discontinuity in the velocity field $\boldsymbol{U}$ at the interface due to the phase change.

Welch and Wilson continue the work of Juric and Tryggvason and develop a method of treating flows with phase change in the framework of the VOF model. They connect the discontinuity of the velocity field $\boldsymbol{U}$ normal to the interface with a discontinuity of the heat flux $\boldsymbol{q}$:

$$\|(\boldsymbol{U} - \boldsymbol{U}_i)\| \cdot \boldsymbol{n} = \left(\frac{1}{\rho_l} - \frac{1}{\rho_g}\right)\frac{\|\boldsymbol{q}\| \cdot \boldsymbol{n}}{\Delta h} \tag{5.57}$$

In equation (5.57), $\|\Psi\|$ denotes the jump of the quantity Ψ at the interface. Furthermore, Δh is the evaporation enthalpy. In this approach only heat conduction is taken into account. Other effects, e. g., the kinetic energy or the viscosity, are neglected.

The temperature is set equal to the saturation temperature in dependence on the pressure at the liquid side:

$$T_i = T_{\text{sat}}(p_l) \tag{5.58}$$

In contrast to equation (5.58), Juric and Tryggvason employ a more complex relation. However, their evaluation of the interface temperature is not without ambiguities.

Welch and Wilson determine the jump of the heat flux from the temperature gradients at both sides of the interface. For this purpose, the interface is reconstructed as a straight line in each cell. The center of the line has the saturation temperature. Once the temperature gradients on both sides of the surface are known, the heat fluxes result by multiplying with the conductivity.

6 Simulations with Star-CD

Figure 6.1 shows what many numerical experiments are concerned with: falling films disturbed by turbulence wires. The final objective was to optimize the evaporators of the EasyMED desalination plant, since the main thermal resistance is assumed to be located there. The dimensions of the seawater film are approximately 0.3 mm × 400 mm × 1500 mm. As is well known, CFD is not able to completely capture a system with so large a difference between magnitudes. So the first step of the CFD analysis was to scale the problem down and to make appropriate simplifications.

Figure 6.1: Film with turbulence wires.

Small, two-dimensional (2D) control volumes with one, two, or three circular obstacles were considered to be best suited to study the influence of the turbulence wires on the film flow. In Figure 6.2, such a system under investigation is depicted. The objective of the 2D numerical experiments was to find out the extension, *L*, of the wire's wake in order to give recommendations for the spacing between the wires. The simulations with set-ups similar to Figure 6.2 were performed with Star-CD [29].

Unfortunately, when the CSF model [25] for the action of a constant surface tension was employed, huge velocities perpendicular to the free surface occurred. With commercial software, it is not possible to lay open the numerics. Obviously, the phenomenon is an error. It is known that the higher the viscosity, the weaker is the effect. Maybe the phenomenon is attributed to "parasitic currents." As a practical consequence, the numerical experiments with Star-CD had to be performed with zero surface tension for laminar cases where the viscosity was small. However, it will be shown

https://doi.org/10.1515/9783110592337-006

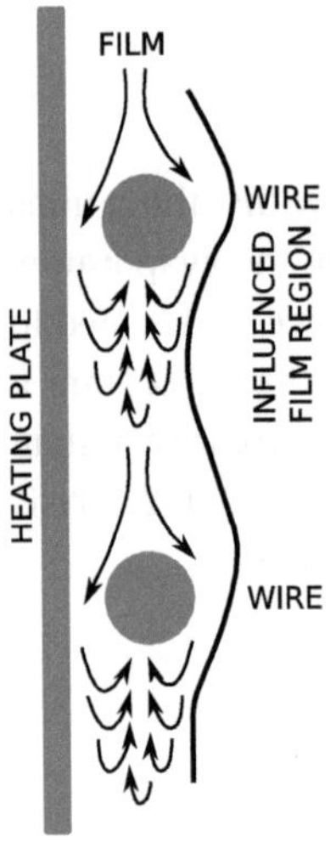

Figure 6.2: Influenced region behind a turbulence wire.

in the course of this monograph that the results of these simulations agree with the findings of OpenFOAM as far as the wire spacing is concerned. (In OpenFOAM, the effect of the surface tension could be included.) So from a pragmatic point of view, the surface tension is not always "necessary." The work presented in this chapter has been published in [116, 117, 115].

6.1 Effect of entrance region

The first step was to study entrance effects. The question was at what distance downstream the film inlet the flow can be regarded as (hydrodynamically) fully developed. While treating this question, the Nusselt velocity profile, equation (4.15), was adopted as a reference. The set-up for the first numerical experiments is depicted in Figure 6.3(top): The film inlet is at the top, with coordinate $x = 0$. In the simulation, one single fluid phase is present. At $y = 0$, there is the plate with the no-slip condition. At $y = \delta$, there is another boundary with zero shear stress. This boundary approximately represents the free surface; at the bottom is the outlet. The flow rate at the inlet is chosen such that further downstream the Nusselt velocity profile for a film of thickness δ can be reached. In one case, a uniform velocity is chosen at the entrance. It can be deduced from this simulation that for water at 25 °C the flow is fully developed after $x = 10$ mm. An analogous numerical experiment was performed with a more realistic parabolic (Poiseuille) inlet velocity profile. Again after $x = 10$ mm, the film flow was found to be fully developed.

For the next step, gas-liquid systems were studied using the volume of fluid (VOF) model [65]. This time the inlet gap was chosen to be 1 mm. For the flow rate, the value

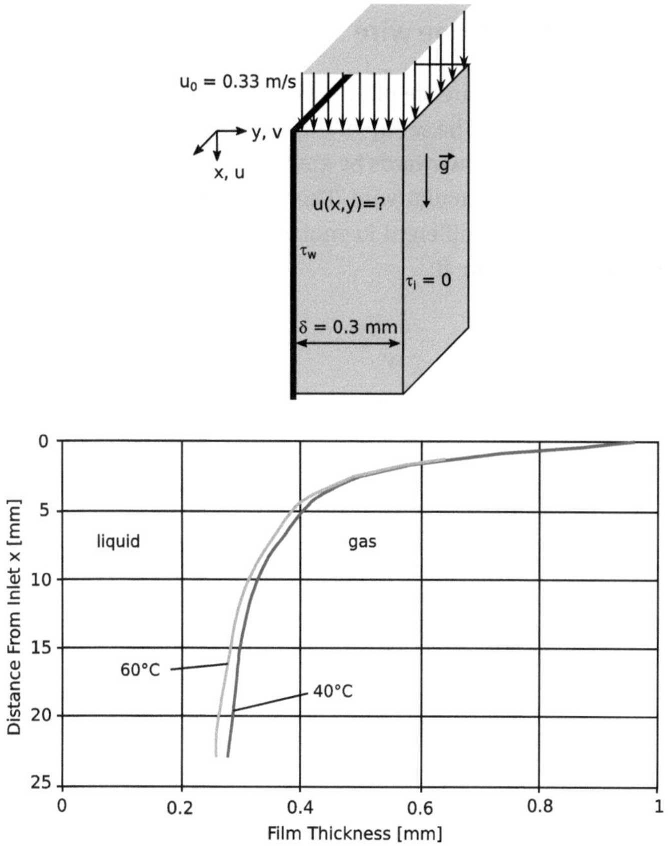

Figure 6.3: Single phase set-up to study entrance effects (top). Film thickness at entrance for two different temperatures according to two phase simulation (bottom).

of 150 l/h was taken, which is typical for the desalination test unit. The result is shown in Figure 6.3 (bottom) for water of 60 °C and 40 °C. It can be seen that the film reaches a thickness of approximately 0.25 mm after an entrance region of 25 mm.

The conclusion from these first numerical experiments is that the region with entrance effects is small and can be neglected. The error introduced by this simplification is insignificant for practical requirements.

In Section 4.3, we already encountered a model of the entrance region. Recall that the velocity profile was always semiparabolic; see equation (4.31) and Figure 4.4. With the aid of equation (4.33), we can calculate the distance x from the inlet where the film thickness $\delta = 1.01\delta_0$. In case of a flow rate of 150 l/h, saltwater at 25 °C, and $\delta_i = 4\delta_0$, the distance is $x = 19$ mm. For a temperature of 40 °C, $x = 23$ mm. For 60 °C, $x = 29$ mm. All of these numbers confirm that the entrance region is short.

6.2 Hydrodynamic studies with one wire

In the hydrodynamic 2D studies, the influence of one wire on the velocity field was examined under isothermal conditions. The set-up of this series of numerical experiments is depicted in Figure 6.4. Forced downwards by gravity, a water film runs along a vertical plate and immerses a single circular wire. The flow rate, Γ, and the water kinematic viscosity, ν, are varied to give different Reynolds' numbers. As Re is so important, let us recall two ways to calculate it:

$$\mathrm{Re} = \frac{\Gamma}{Z \cdot \nu} = \frac{\delta \cdot \overline{u}}{\nu}, \tag{6.1}$$

where $Z = 400$ mm is the film width, δ the initial film thickness, and $\overline{u}$ the average inlet water velocity. The values of Re were 50 and 150 for laminar flow, and 1000 and 3000 for the fully turbulent regime. The film thickness, δ, and the kinematic viscosity, ν, chosen for the numerical experiments are listed in Table 6.1. Beside of Re, the wire diameter, d, was also varied in dependence on the film thickness, δ, of the undisturbed film. The quantity d was chosen to be $d = \delta/2, \delta, 3\delta/2$.

The distance between plate and wire was kept constant, $d^* = 0.2$ mm; cf. Figure 6.4. The numerical grid is defined such that the film is at least 20 cells thick. The control volume's height is of the order of 4 cm. The gas inside the control volume is mainly at rest.

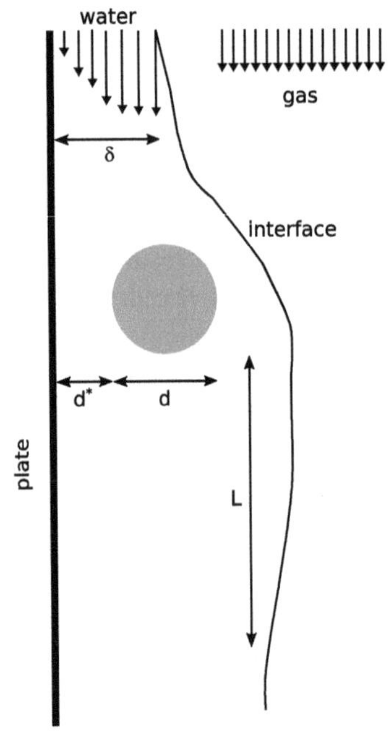

Figure 6.4: Set-up of hydrodynamic numerical experiments with one wire.

Table 6.1: Film thickness and kinematic viscosity.

Re	δ [mm]	ν [10^{-6} m^2/s]
50	0.25	1
150	0.36	1
1000	0.45	0.475
1500	0.65	0.475

Special care was taken for the simulation of turbulent film flow, for which a low Reynolds' number k-ε model was employed; cf. Section 3.4.3 and [88]. In a larger control volume, a film was simulated that was not disturbed by a wire. At the inlet, the default value for the turbulent intensity (2.5 %) and the mixing length (inlet extension/10) were taken. Near the outlet, the flow field was assumed to have reached a stationary state. The values at the outlet for the velocity, turbulent energy and energy dissipation rate were fitted by analytical functions and taken as inlet conditions for the simulations with wire. One distinct result was that a turbulent film is thicker than a laminar one. For Re = 1000, the film thickness increased from 0.425 m at the inlet to 0.45 mm at the outlet. In the case of Re = 3000, the values were 0.6 mm and 0.65 mm, respectively.

For the simulation of the Laplace pressure, the continuum surface force (CSF) model [25] is optional within Star-CD. As mentioned earlier, it produced numerical errors for thin water films. This effect decreases with increasing viscosity. Therefore, the CSF model had to be switched off for Re = 50 and 150, whereas it served well at higher Reynolds' numbers, where the turbulence caused large viscosity values.

The aim of the numerical experiments was to determine the length L of the wire's wake which should be the approximate optimal distance between the wires in the grid of the desalination plant. The obtained results are summarized in Table 6.2. The uncertainty of the values given for L/d is rather large since the transition from disturbed to undisturbed film flow is smooth. The uncertainty is estimated to be about 30 %. It can be seen from Table 6.2 that the differences between laminar and turbulent film flow are distinct. The values of L/d in the turbulent regime are smaller by a factor of

Table 6.2: Wake extension for different Re and d.

	d/δ		
	0.5	1.0	1.5
Re	L/d		
50	20	18	14
150	30	20	20
1000	12	12	12
3000	12	7	–

about 2 in comparison to the laminar flow. The turbulence has a strong mixing effect on the flow field such that the perturbation of the wire upstream is sooner "forgotten." The missing value in Table 6.2 could not be obtained because for such a big wire diameter the film flow was waterfall-like and did not converge to an immersing flow. The conclusion drawn out of Table 6.2 for the wire spacing in the turbulent regime is

$$L/d = 12. \tag{6.2}$$

In the case of laminar film flow, the situation is more complicated. The following ansatz is made:

$$L/d = f(\mathrm{Re}, d^3 g/\nu^2), \tag{6.3}$$

where g denotes the gravitational acceleration. The second term resembles the Nusselt expression for the Reynolds' number

$$\mathrm{Re} = \frac{\delta^3 g}{3\nu^2}. \tag{6.4}$$

In the laminar case, the following correlation was obtained for the wire spacing:

$$L/d = 22 \cdot \left(\frac{\mathrm{Re} \cdot \nu^2}{d^3 \cdot g}\right)^{0.09} \tag{6.5}$$

A comparison of the above correlation with the values determined in the numerical experiments is shown in Table 6.3. There are only 6 values to be fitted, so it is evident that other relations than equation (6.5) are possible, in principle. However, equation (6.5) accounts for the fact that the determined values for L/d are very similar.

Table 6.3: Comparison between Table 6.2 and correlation equation (6.5).

	d/δ		
	0.5	1.0	1.5
Re	L/d		
50	20 (24)	18 (20)	14 (18)
150	30 (24)	20 (20)	20 (18)

6.3 Wake of a wire

The wake of a wire shall be given more attention. On the left-hand side of Figure 6.5, the velocities, simulated by Star-CD, can be seen. The temperature of the water is 20 °C.

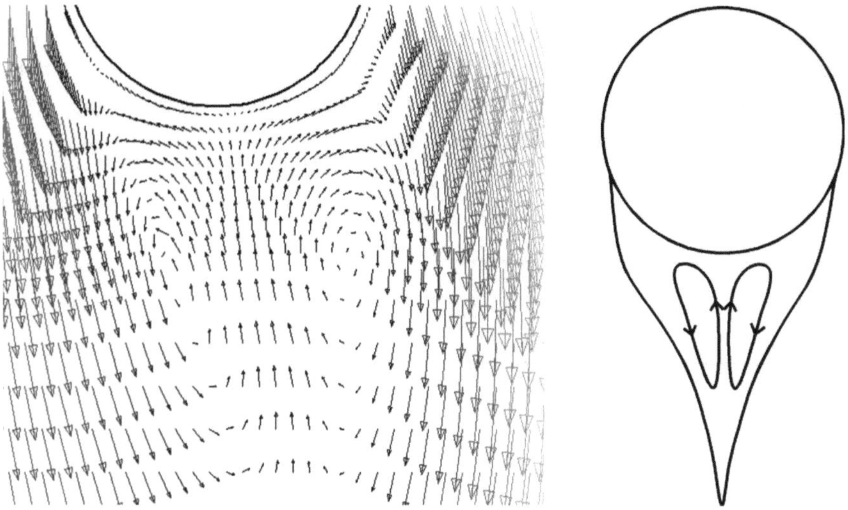

Figure 6.5: Recirculation in the wake of a wire.

The Reynolds' number has a value of 150. The diameter of the wire is equal to the thickness of the corresponding undisturbed film. In the wake of the wire, there are apparently two zones of recirculation. The left eddy rotates anti-clockwise, the right one clockwise. On the right-hand side of Figure 6.5, this peculiarity of the flow is drawn more clearly. If there are more wires to disturb the falling film, their distance to each other should be large enough, so the next wire is located outside the region of recirculation.

6.4 Thermal studies

The results delivered in Section 6.2 were preliminary since the effect of the wire on the temperature field inside the film had been neglected so far. As a first further step, assumptions had to be made on the temperature of the heating plate and of the vapor. Since the temperature drop between neighboring effects of the desalination unit is approximately 5 °C, this difference was assumed between the heating plate and the vapor. At the inlet, the water temperature was chosen to be equal to the vapor temperature. First, the thermal studies were performed with only one wire. These results served as the basis for the calculations with two wires.

In the thermal numerical experiments with one wire, two temperatures were selected that can be considered as extremes in the evaporation cells of the desalination plant. Simulations were performed with inlet temperatures of 40 °C and 60 °C. Flow rates were chosen to give the Reynolds' numbers of the laminar and the turbulent regime. In this series of numerical experiments, the distance between the heating plate and the wire was $d^* = 0.1$ mm which is to be compared to the value of 0.2 mm in

Section 6.2. This change was necessary in order to have the wire immersed by the film for lower viscosities at laminar Reynolds' numbers. Since the actual value for d^* in the desalination plant is unknown, the change in the geometry has also the advantage to get another flow field in comparison to Section 6.2.

Table 6.4: Wake extension L deduced from thermal studies.

			d/δ	
		Flow rate	0.5	1.0
Re	Pr	[l/h]	L/d	
160	4.12	150	24	28
220	2.87	150	24	28
720	4.12	680	30	25
1000	2.87	680	30	27

The results for the extension L of the wire's wake in units of the wire diameter d are summarized in Table 6.4. The values of L/d are very similar for different temperatures as well as for laminar and turbulent films. The conclusion from these numerical experiments for the distance between the wires is

$$L/d = 27. \tag{6.6}$$

The perturbation of the velocity and of the temperature field downstream the wire were monitored to obtain the results in Table 6.4. Not much of a difference between these flow fields could be found. The transition to an undisturbed flow is again smooth and its uncertainty was estimated to be about 30 %.

The disturbance of the film flow by the first wire was observed experimentally to be rather weak, while a significant enhancement of turbulence establishes when using a multitude of wires; cf. Figure 6.6. To improve the accuracy, at least one more wire had to be taken into consideration. So the next step was to simulate the film flow around two wires. The chosen Reynolds' number was Re = 220, corresponding to a water flow rate of 150 l/h at 60 °C. Under these conditions, the thickness of the undisturbed film is approximately $\delta = 0.25$ mm. The distance between the neighboring wires was varied: It was chosen to be $0.67L_1$, L_1, and $1.33L_1$, where L_1 is the recommended wire distance from the previous simulation with only one wire. The wire diameter was $d = \delta/2$ and $d = \delta$. Note that in simulations with single wire the recommended distances were $L_1 = 24d$ for $d = \delta/2$ and $L_1 = 28d$ for $d = \delta$.

In the simulations with two wires of diameter $d = \delta/2$, the velocity as well as the temperature profiles in the wake of the second wire obtained at different spacings were found to be almost identical. For the wire diameter $d = \delta$ at spacing of $0.67L_1$ and L_1, there is no strong difference between the disturbances of the film in the region $L_2 = 28d$ behind the second wire. However, when approaching the second wire from

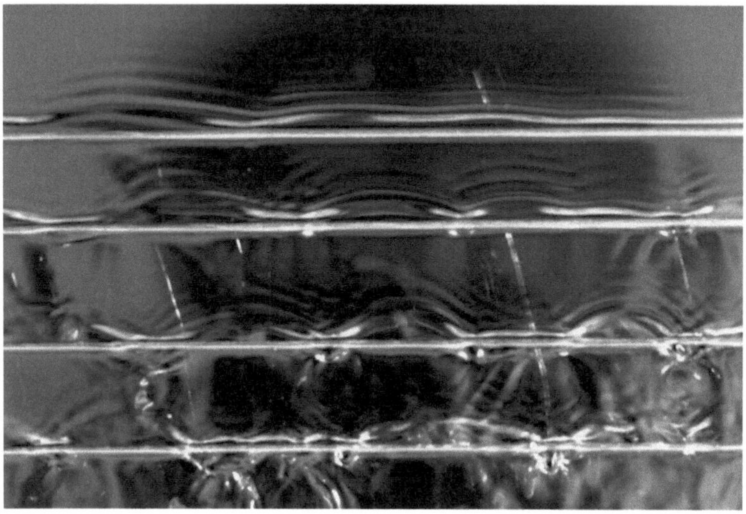

Figure 6.6: A multitude of wires causes turbulence.

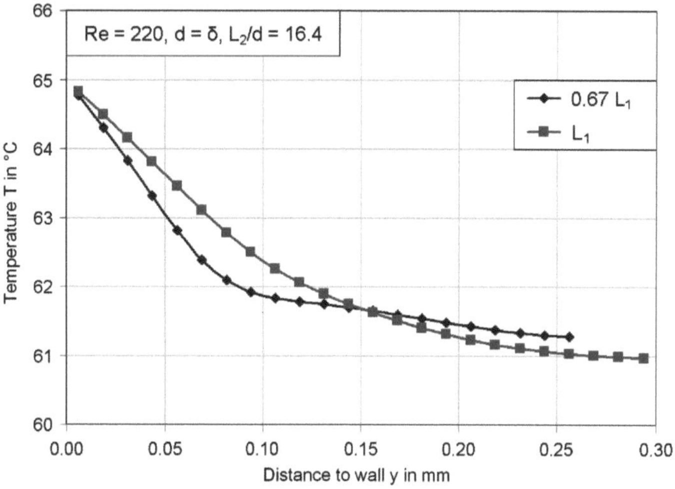

Figure 6.7: The smaller wire spacing results in a higher heat transfer.

below, the picture changes. Figure 6.7 shows a comparison of the temperature profiles at a distance $L_2 = 16.4d$ below the second wire. In the case of a smaller wire spacing, the temperature field is much more disturbed. Notice the steeper slope at $y = 0$ which indicates a higher heat transfer from the heating plate.

The conclusion from these numerical experiments is that the recommended distances between the wires, based on observations with one wire (Table 6.4), have been estimated too large. A reduction of this quantity by a factor of approximately 0.67 deduced from the thermal analysis with two wires is appropriate. On average, the new recommendation for the wire spacing is

$$\boxed{L/d = 18}. \tag{6.7}$$

Note that the above equation has been obtained from simulations without surface tension. Years later it was confirmed by further numerical experiments with surface tension.

6.5 New turbulence model

An extension of the k-ε model has been proposed and described in detail in Section 3.4.5 of Chapter 3. A sink term S_k in the equation for $\bar{D}k/\bar{D}t$ and a source term S_ε in the equation for $\bar{D}\varepsilon/\bar{D}t$ were postulated:

$$\begin{aligned}\frac{\bar{D}k}{\bar{D}t} &= \Sigma_k - S_k \\ &= \Sigma_k - C_{\sigma 1}\frac{\sigma/\xi}{\rho_L t_\xi} \\ &= \Sigma_k - c_{\mathrm{VOF}} a_k\left(1 - \exp\left(-b_k\left(\frac{y}{\delta}\right)^{c_k}\right)\right)\frac{\sigma\varepsilon^2}{\rho_L k^{5/2}},\end{aligned} \tag{6.8}$$

$$\begin{aligned}\frac{\bar{D}\varepsilon}{\bar{D}t} &= \Sigma_\varepsilon + S_\varepsilon \\ &= \Sigma_\varepsilon + C_{\sigma 2}\frac{\sigma/\eta}{\rho_L t_\eta^2} \\ &= \Sigma_\varepsilon + c_{\mathrm{VOF}} a_\varepsilon\left(1 - \exp\left(-b_\varepsilon\left(\frac{y}{\delta}\right)^{c_\varepsilon}\right)\right)\frac{\sigma\varepsilon^{5/4}}{\rho_L \nu_L^{7/4}},\end{aligned} \tag{6.9}$$

where Σ_k and Σ_ε represent the terms of the original k-ε model. In order to predict a certain turbulent film flow, values for the coefficients a, b, and c in the equations (6.8) and (6.9) must be known.

The new turbulence model has been implemented into Star-CD and a multitude of numerical experiments were performed to find out in which order of magnitude the closure coefficients have to be. A Reynolds' number of $\mathrm{Re} = 1000$ was chosen for the fully turbulent film flow. The first simulations were unstable. After a few time steps, the turbulent kinetic energy k inside the film approached zero thus resulting in a huge sink term S_k (cf. equation (6.8)) and making the calculation divergent. As a consequence, the coefficient a was gradually decreased. Finally, at a very small value for a, the simulation was stable.

Figure 6.8 shows the turbulent viscosity μ_t inside the film for a calculation with the new turbulence model and by the common equations of hydrodynamics. It is clearly

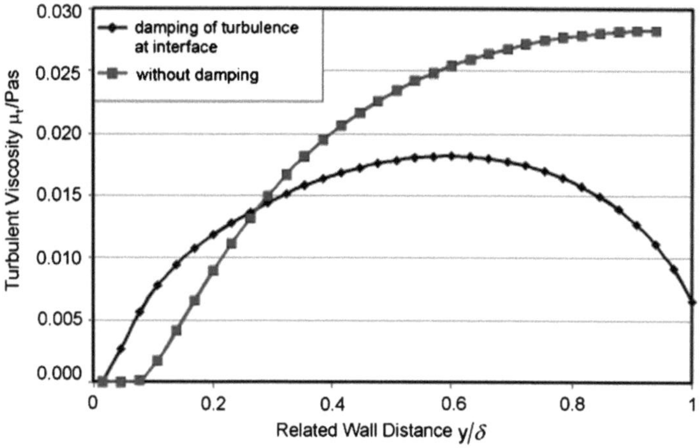

Figure 6.8: The turbulence is damped near the free surface on account of the new model.

seen that the new model causes a damping of the turbulence at the interface. The chosen values for the coefficients a, b, and c were

$$a = a_k = a_\varepsilon = 1.25 \cdot 10^{-45}; \quad b = b_k = b_\varepsilon = 1; \quad c = c_k = c_\varepsilon = 2. \tag{6.10}$$

The coefficient a has to be so small to counteract against numerical instabilities. Note that in the equation for k (6.8) the sink term depends on $k^{-5/2}$. Thus with the reduction of k the sink term becomes larger and larger. In a similar manner in the equation for ε (6.9), the source term contains $\varepsilon^{5/4}$. With the progression in time, the source term becomes more and more dominant. It requires further major efforts to make the new model a realistic description of turbulent film flow. However, the feasibility of it is herewith demonstrated.

7 Employment of OpenFOAM

In this chapter, it will be shown that OpenFOAM [2] is well suited for the direct numerical simulation of laminar wavy falling films. As the name suggests, it is a free, open source CFD software which can be downloaded on the Internet. For our purpose, the application *interFoam* is employed. How this can be done is detailed in Annex B. Since the source code is laid open, it can be the basis for further programming. Thus the new application *thinter* was developed. Its name means *thermal interFoam*. For instructions, how to program it, read Annex C.

First, the hydrodynamical side is tested by two-dimensional simulations with *interFoam*, Section 7.1. Three-dimensional numerical experiments are still computationally very expensive. However, they are feasible as it is demonstrated in Section 7.2. OpenFOAM can be run on several cluster nodes in parallel with the aid of the message passing interface (MPI). In Section 7.3, peculiarities of wavy falling films are collected. Also results from *thinter* are shown there. Finally, in Section 7.4, the film flow is disturbed by wires to check the recommendation obtained with Star-CD. These simulations are thermal. In Annex D, details are given, how the mesh around turbulence wires can be defined. Throughout this chapter, the surface tension is always nonvanishing.

Earlier publications on our research with OpenFOAM are [119, 118, 131, 132]. The short communication [119] might have been too short, since there were misunderstandings. They are corrected in Annex E.

7.1 2D periodically excited waves

OpenFOAM was thoroughly tested before numerical experiments with turbulence wires were undertaken. The experiments by Nosoko et al. [105] on two-dimensional (2D) excited waves in laminar falling films served as a test case. The correlations for the phase velocity, equation (4.66), and for the maximal peak height, equation (4.67), were compared with the simulation. Water at 20 °C for falling films at Re = 25, 40, 60 were employed to be as close to the experiments as possible. At the inlet, the velocity u was varied with time t according to

$$u = u_0(1 + 0.03\sin(2\pi ft)), \tag{7.1}$$

where u_0 is the average velocity of a smooth film with the thickness $\delta_0 = \delta_0(\mathrm{Re})$ and f is the frequency of the perturbation. It was chosen 13 Hz, 20 Hz, and 45 Hz. The cell size was $(\Delta x, \Delta y) = (0.4\delta_0, 0.1\delta_0)$ following the example of Kunugi and Kino [81], that is, the average film thickness was approximately ten cells.

OpenFOAM 1.1 was not mature enough to simulate falling films. First, numerical experiments with this earlier version showed instabilities leading to a dry spot on the

https://doi.org/10.1515/9783110592337-007

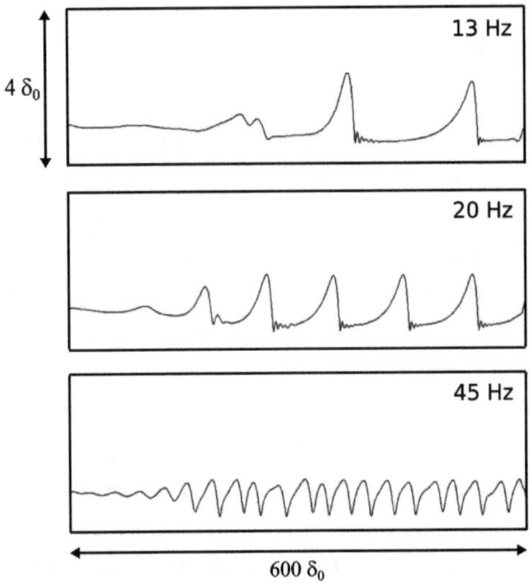

Figure 7.1: Falling water films at Re = 60 excited with a frequency $f = 13$ Hz (top), $f = 20$ Hz (middle), and $f = 45$ Hz (bottom).

wall. In such a case, the program set the time step to zero, that is, the simulation did not proceed any more. Therefore, we upgraded to the OpenFOAM version 1.3 where the simulation of falling films was much more stable.

Figure 7.1 shows three periodically excited waves at Re = 60 with $f = 13$ Hz (top), $f = 20$ Hz (middle), and $f = 45$ Hz (bottom). The pictures have been rotated to better fit the page. In this representation, the flow is from the left to the right. Near the inlet, the perturbation is very small. However, further downstream large solitary waves have developed. In front of the solitary waves, there are small capillary waves. With increasing frequency, the wave pattern becomes more and more sinusoidal.

Figure 7.2 compares the wave phase velocity and wave peak height obtained from the experimental correlations (horizontal axis) with the values from the simulation (vertical axis). The diagonal line of unit slope marks perfect agreement. Error bars are also displayed in Figure 7.2. One source for the error bars is the 0.5 mm uncertainty of the ruler by which distances were measured from the screen. This error propagated according to Gauss [26] to give the uncertainty in the appropriate units. If the variation from wave to wave could not be explained by the ruler error, an additional uncertainty has been added quadratically. As can be captured from Figure 7.2, the agreement between experiment and simulation is very good. Thus it can be concluded that already OpenFOAM 1.3 was well suited to simulate free surfaces under the influence of the Laplace force in the case of falling water films.

The result of a grid refinement test is displayed in Figure 7.3. (This time an OpenFOAM version of the year 2018 was used. It ran under the Windows 10 operating system.) The surface of the falling film after a time of 0.2 s is shown. The Reynolds' num-

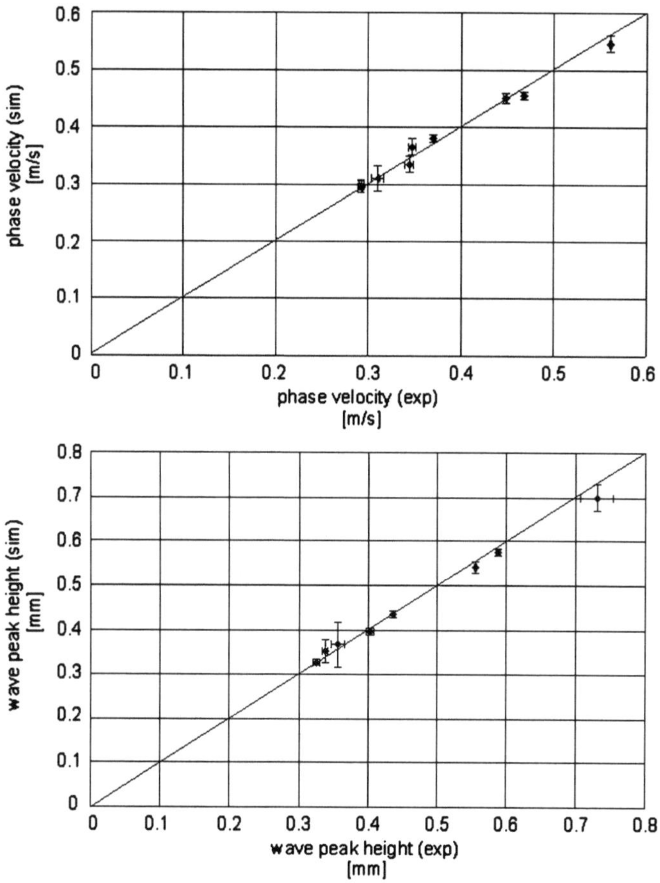

Figure 7.2: Comparison of the wave phase velocities (top) and wave peak heights (bottom) obtained experimentally (horizontal axis) and numerically (vertical axis).

ber is 60, the temperature 60 °C, and the excitation frequency 20 Hz. The numbers 2, 1, 1/2, and 1/4 indicate the refinement factor. The factor 1 refers to the default cell size $(\Delta x, \Delta y) = (0.4\delta_0, 0.1\delta_0)$. A factor of 2 means a coarsening: $(\Delta x, \Delta y) = (0.8\delta_0, 0.2\delta_0)$. The simulation with the coarser grid is obviously different. The refinement factors 1/2 and 1/4 yield almost identical results. Convergence can be observed. With the default cell size the simulation is slightly different from the next refinement step. However, it is satisfactorily close.

7.2 3D simulations

OpenFOAM also performed calculations in parallel employing the message passing interface (MPI) [58, 108]. The 3D simulations took place at the Paderborn Center for

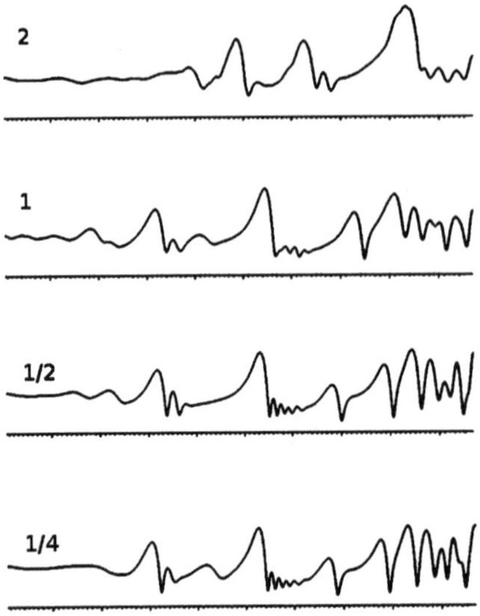

Figure 7.3: Grid refinement test. The number in each plot refers to the refinement factor.

Parallel Computing (PC^2). In order to capture a 3D falling film, a coarse numerical grid was used: $(\Delta x, \Delta y, \Delta z) = (0.6\delta_0, 0.2\delta_0, 1.0\delta_0)$, where δ_0 denotes the thickness of a smooth film with the same Reynolds' number. Note that approximately on average the film was only five cells thick. We followed again the example of Kunugi and Kino [81]. The numerical grid consisted of about 4.5 million cells.

Re = 75 was chosen since from this Reynolds' number the waves become unstable. The wall had a constant temperature of 65 °C and the water was 60 °C warm at the inlet. The OpenFOAM application *interFoam* was extended. At that time, this extension was named *thinterfoam*.

Figure 7.4 shows a black and white representation of the falling film surface. White marks the wave crests and black the wave troughs. It can be seen that two-dimensional

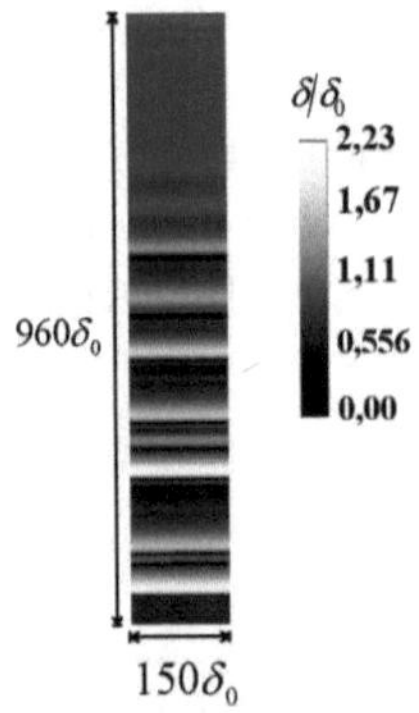

Figure 7.4: Black and white representation of the surface of a 3D falling film at Re = 75.

waves have developed although no excitation had taken place at the inlet. This phenomenon is due to numerical and hydrodynamic instabilities. Near the bottom traces of a transition to three-dimensionality can be spotted.

7.3 Peculiarities of wavy falling films

In this section, some figures shall be displayed and explained that give insight into the nature of laminar wavy falling films. We shall be first concerned with the hydrodynamics and then proceed to the heat transfer.

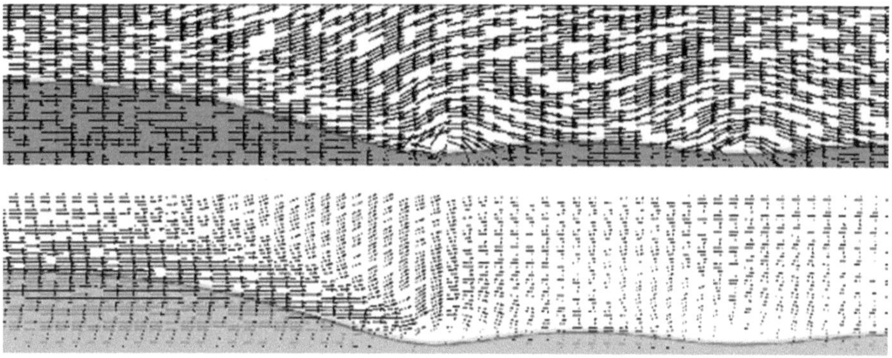

Figure 7.5: Velocity vector plots: Direction of velocity (top), direction and magnitude of velocity (bottom). The solitary wave is the large wave on the left. The capillary waves are the small waves on the right.

Figure 7.5 shows velocity vector plots. In the upper picture, only the velocity direction is given, that is, all the arrows have the same length. In the lower figure, the arrow length is a measure for the velocity magnitude. It can be seen from Figure 7.5 that in the first two troughs of the capillary waves in front of the solitary wave the velocity is very small and the flow recirculating. Thus the simulation nicely reproduces the experimental observations by Tihon [136]. Dietze devoted his Ph. D. thesis to this phenomenon [38].

The upper part of Figure 7.6 displays the temperature in a falling film heated by the wall. For the same film at the same instant, the thickness is shown in the lower part of Figure 7.6. The transition zone between water and air is so large since the 3D numerical grid was rather coarse. (In Figure 7.6 (bottom), the volume of fluid (VOF) scalar c_{VOF} is displayed.) In general, the thinner the film the better is the heat transfer. Furthermore, further downstream the average temperature should be higher than it is upstream. These two statements hold true also in the case of a wavy falling film. However, the wave dynamics are so complicated that an experiment or a simulation are needed to predict the exact temperature field.

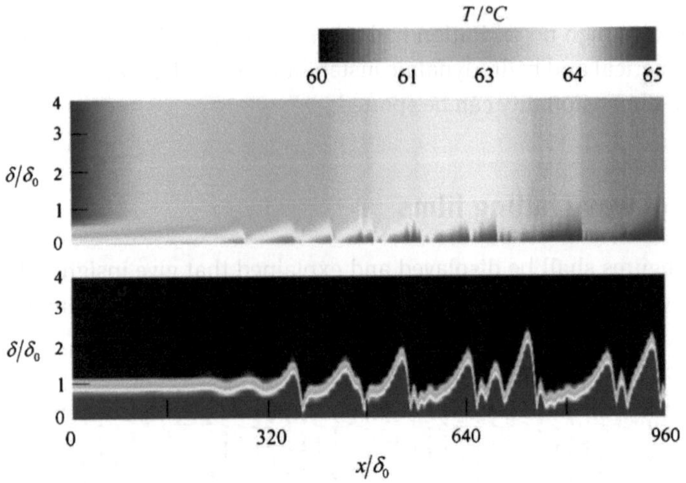

Figure 7.6: Temperature field in a falling film heated by the wall (top) and corresponding film thickness (bottom).

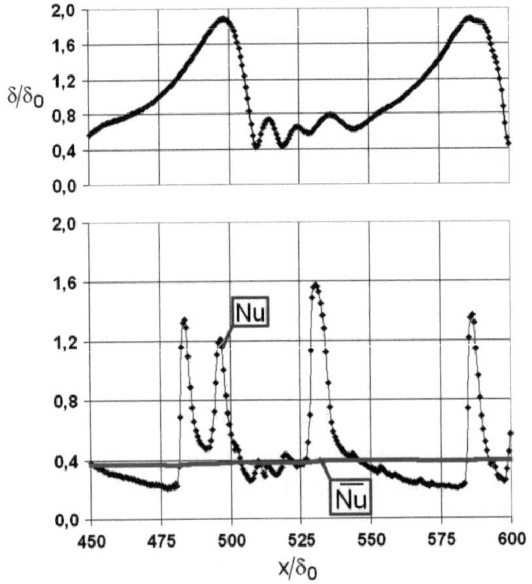

Figure 7.7: Local and average Nusselt number (bottom) in comparison to film thickness (top) for a falling film with Re = 75 and constant wall temperature.

Also the local Nusselt number, Nu, is difficult to predict on account of the complicated wave dynamics. Figure 7.7 (bottom) shows Nu in comparison to the film thickness, Figure 7.7 (top). The mean Nusselt number, $\overline{\text{Nu}}$, is displayed, too. Its value of 0.4 can be compared to the prediction of the VDI Heat Atlas, equation (4.55) [142]: 0.3 for Re = 75 and constant wall temperature.

How is the Nusselt number calculated in this specific case? From equation (4.30), it is known that

$$\mathrm{Nu} = \frac{\alpha_x}{\lambda_L}\left(\frac{\nu_L^2}{g}\right)^{1/3}. \tag{7.2}$$

The local heat transfer coefficient, α_x, is defined by

$$q = \alpha_x \cdot (T_W - T_I), \tag{7.3}$$

where q denotes the heat flux, T_W the wall temperature, and T_I the temperature of the gas-liquid interface. The heat flux can be determined via

$$q = -\lambda_L \left.\frac{\partial T}{\partial y}\right|_W \approx \lambda_L \frac{T_W - T_{\text{cell, next to wall}}}{0.5\Delta y}. \tag{7.4}$$

In order to find the mean Nusselt number, $\overline{\mathrm{Nu}}$, the local heat transfer coefficient is averaged along the streamwise coordinate x.

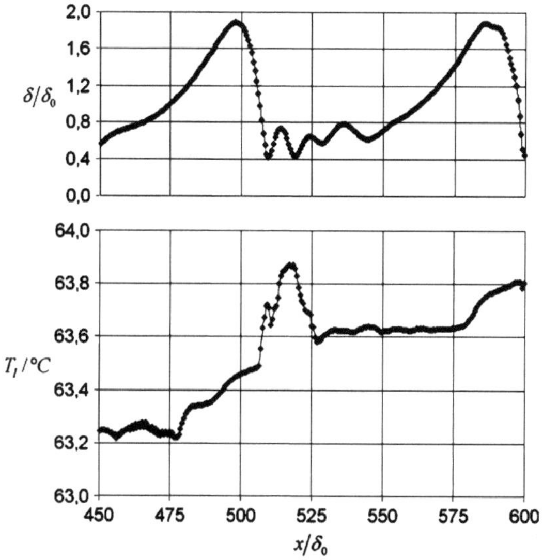

Figure 7.8: Temperature of the gas-liquid interface (bottom) in comparison to the film thickness (top) for a falling film with Re = 75 and constant wall temperature.

Figure 7.8(bottom) shows the temperature of the gas-liquid interface, T_I, for a vertically falling film of Re = 75. On the top, the film thickness related to the thickness of the smooth film is displayed. The highest T_I can be found in the region, where the film is thin. This observation corresponds to what we know on a smooth film, Section 4.2. The smaller the film thickness the higher is the local heat transfer coefficient, equation (4.27). It can also be seen from Figure 7.8 (bottom) that the mean (surface) temperature is rising further downstream.

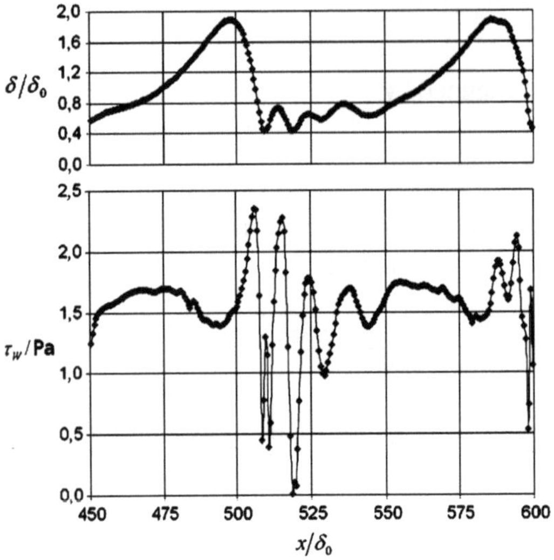

Figure 7.9: Wall shear stress (bottom) in comparison to film thickness (top) for a falling film with Re = 75.

In Figure 7.9 (bottom), the shear stress at the wall is displayed. Again the related film thickness is shown at the top. The wall shear stress is calculated from

$$\tau_w = \mu_L \left.\frac{\partial u}{\partial y}\right|_w, \tag{7.5}$$

where μ_L denotes the dynamic viscosity of the liquid. From equation (4.34), we know that for a smooth film

$$\tau_{w,\,\text{smooth film}} = \rho_L g \delta. \tag{7.6}$$

Note that in regions of backflow, like in the troughs of the capillary waves, the wall shear stress could be even negative. From the figure, we see that in these zones the wall shear stress clearly has minima. However, negative values of τ_w cannot be perceived in this example.

Four different views of the velocity inside a roll wave are displayed in Figure 7.10. The set-up of the numerical experiment is the same as in Annex B, that is, the Reynolds' number is 60, the temperature has a value of 62 °C, and the excitation frequency is 20 Hz. In the picture at the top, isolines of the velocity magnitude U in

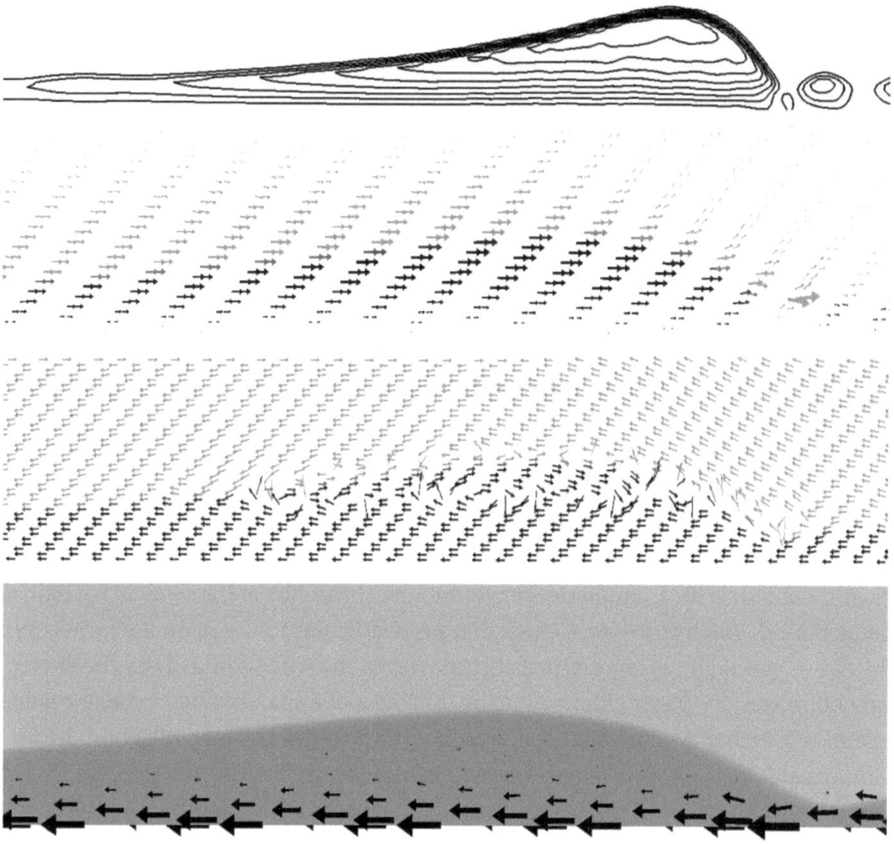

Figure 7.10: Four different views of the velocity inside a roll wave.

the laboratory frame are shown. A single vortex in the solitary wave is apparent. This is a good sign, see also Chapter 8. Dietze, [38] pages 58 and 59, sees one vortex in the roll wave, too. (His reference frame is that of the wave celerity.) The experiments by Portalski [111, 112] and Alekseenko et al. [12] confirm the formation of a single eddy. In Figure 7.10 (top), the vortex of the neighboring capillary wave can also be seen.

In the next plot (second from top), the vectors of the velocity $\boldsymbol{U}$ in the laboratory frame clearly point in the streamwise direction. Not all cells are shown, only a sample. The black colored arrows mark the region of the film, where c_{VOF} is one. The velocity is, to a large extent, unidirectional which justifies the approximation in Section 4.13. The two lower plots (third and fourth from top) are vector representations of the relative velocity inside the film, calculated by

$$c_{\text{VOF}}\left(\boldsymbol{U} - 0.4\,\frac{\text{m}}{\text{s}}\boldsymbol{i}\right).$$

Recall that $\boldsymbol{i}$ is the unit vector in the x direction. In the third picture from top, all arrows have the same length; they only show the direction. Near the wall, these arrows all point backwards, near the surface they point in the streamwise direction. That is the reason why the hump is called a roll wave. The critical reader may spot several vortices. However, in this special reference frame, the velocity is very small in the interior of the solitary wave which is confirmed in the vector plot at the bottom of Figure 7.10. Here, the vector length marks the magnitude of the relative velocity. Again, only a cell sample has been selected.

7.4 Numerical experiments with wires

After OpenFOAM 1.3 has been assessed for the simulation of wavy falling films, it could be employed to answer the question what the optimal distance between turbulence wires is. Figure 7.11 shows the geometry of a single wire. Since dry patches at the wall had to be avoided, the wire was attached to the wall. For the sake of simplicity, we chose a semicircle for 2D numerical experiments. The radius of the semicircle shall be denoted by d which is the wire extension perpendicular to the main flow direction. The radius should be as large as possible. However, the wire should always be wetted. Our choice was $d = 0.8\,\delta_0$, where δ_0 is the thickness of a smooth film having the same Reynolds' number. As indicated in Figure 7.11, the distance between inlet and the first wire was always $50\delta_0$. This length was sufficient to let the falling film develop hydrodynamically. The total length of the control volume was $600\delta_0$ and its height $4\delta_0$.

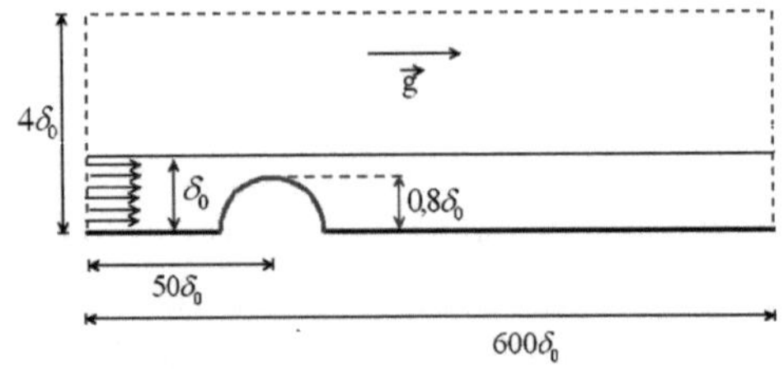

Figure 7.11: Geometry of numerical experiments with wires performed by OpenFOAM; only the first wire is displayed.

In order to decide at which instances the falling film should be observed, three alternatives were considered (compare with Figure 7.13):

1. After a time $t = 0.15$ s, the perturbation of the wires has reached the outlet.
2. After a time $t = 0.2$ s, the waves were fully developed.
3. The time $t = 0.5$ s was considered as a "long time" after which no new phenomena were to be expected.

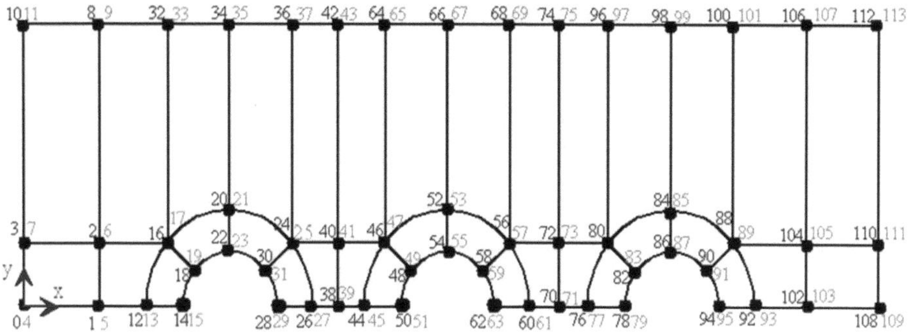

Figure 7.12: Blocks of the grid around three wires.

It was found that the differences from case to case for $t = 0.15\,\mathrm{s}$ and $t = 0.2\,\mathrm{s}$ were not much pronounced. So $t = 0.5\,\mathrm{s}$ was regarded to give the clearest distinctions.

The mean surface temperature was chosen as a measure for the heat transfer. However, around the wires the numerical mesh was block-structured. So the size and orientation could change from cell to cell in this region. Therefore, without loss of significance, the mean surface temperature was determined further downstream where the cells were rectangular.

First, numerous numerical experiments with two wires were undertaken. The Reynolds' numbers were $\mathrm{Re} = 37, 73, 147, 220$. The distance between the wires depended on the wire radius d. Spacings of $10d$, $15d$, and $20d$ were chosen. The water at the inlet had a temperature of 60 °C and the wall with the wires had a temperature of 65 °C. However, the simulations with two wires were not conclusive.

Therefore, the next step was to employ control volumes with three wires. See Figure 7.12. In the beginning, the same temperature boundary condition was applied to the wires as in the cases with two wires, i. e., wall and wires had the same constant temperature of 65 °C. The Reynolds' numbers were $\mathrm{Re} = 73, 147, 220$. No falling films with $\mathrm{Re} = 37$ were simulated since the cell sizes would be very small which would require a simulation time of about 10 days, whereas for the other Reynolds' numbers it took only about half the time. Spacings between the wires of $5d$, $10d$, $15d$, $20d$, and $25d$ have been considered. Furthermore, for $\mathrm{Re} = 73, 147$, adiabatic wires were employed, which means that the temperature gradient had to vanish at these boundaries. The wire distances were again $5d$, $10d$, $15d$, $20d$, and $25d$.

There were two difficulties that occurred in very few cases. First, in the case of a narrow wire spacing the crest of a wire might not be covered with water. Since the simulation with surface tension did not proceed, the surface tension was set to zero for a very short time interval and reset to its initial value afterwards. Sometimes a huge wave developed from this workaround. However, after $t = 0.5\,\mathrm{s}$ the huge wave had disappeared through the outlet. The other difficulty took place in cases with adiabatic wires. After some time the temperature diverged. (The possibility to reduce the

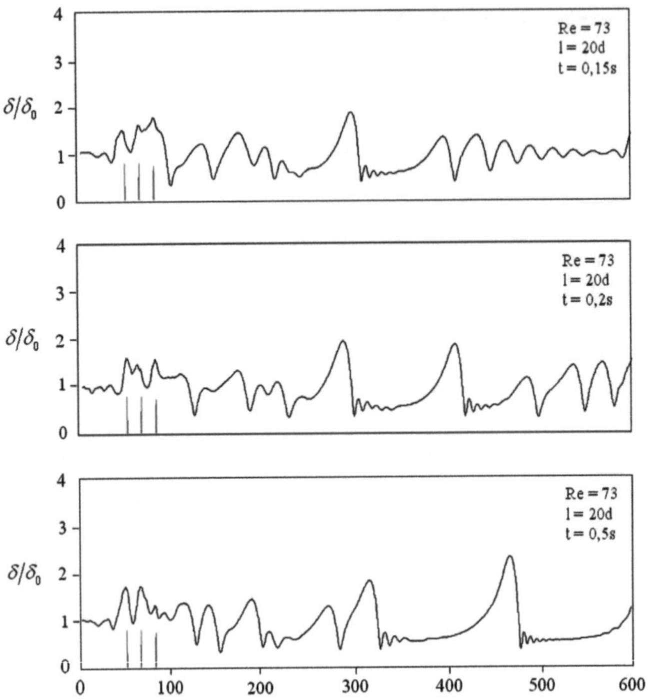

Figure 7.13: The interfaces of a falling film at Re = 73 at the times 0.15 s (top), 0.2 s (middle), and 0.5 s (bottom). The distance between the wires is 20*d*.

`writeInterval` was not taken into consideration, at that time.) These few cases could not be taken into account for the evaluation of the heat transfer after $t = 0.5$ s.

Figure 7.14 shows the surfaces of three falling films with a wire spacing of 5*d* (top), 10*d* (middle), and 15*d* (bottom) at Re = 73. In general, it has been observed that the stronger the perturbation by the wires, the wavier the falling film and the higher the mean surface temperature. From this, we concluded a better heat transfer and a higher evaporation rate. For the cases in Figure 7.14, it is straightforward to see that the wire distance of 15*d* gives a better heat transfer than 5*d* or 10*d*.

It has been observed after $t = 0.5$ s *in all the cases with three wires* that a wire spacing of 15*d* or 20*d* gave the best results. *Thus, with all due caution this result is regarded as a confirmation of the older result of* 18*d as optimal wire distance.*

In Figure 7.15, the wake of a wire is displayed. Only the directions are shown by the arrows. The darker grey marks the water, the lighter grey is the air. The picture is from an interFoam simulation with Re = 220, $T = 60$ °C, $f = 0$ Hz, and $t = 0.42$ s. The first wire of three is shown. How such a case can be realized, is described in Annex D. In Section 4.12, the zero streamline was introduced. From Figure 7.15, it can be seen that this streamline does not reach far. Its extension is in the order of magnitude of 2*d*.

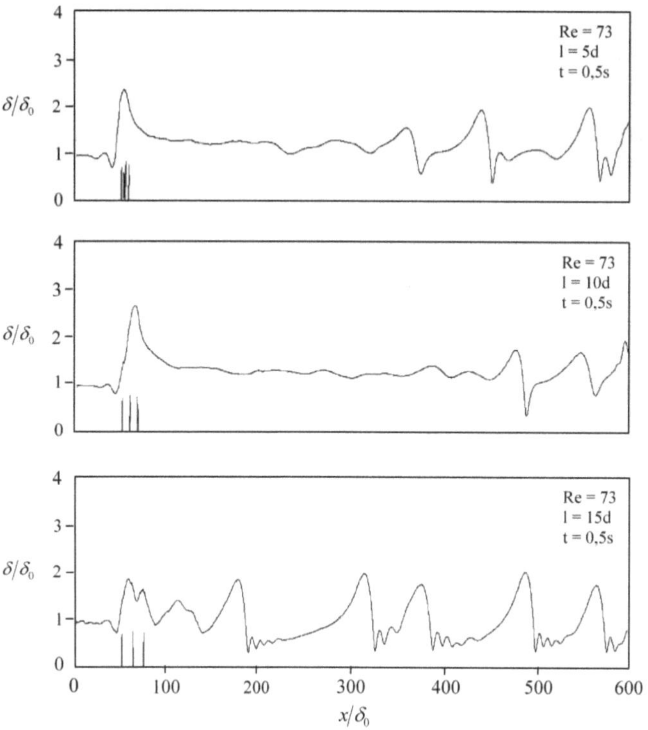

Figure 7.14: The interfaces of three falling films at Re = 73 with a wire spacing of 5*d* (top), 10*d* (middle), and 15*d* (bottom).

Some helpful details on how to program the mean surface temperature are mentioned now. First, get the coordinates of the cell centers. For this purpose, go into the directory with your case. For this example, it has the name `casename`. Type:

```
postProcess -func writeCellCentres -case casename -time 0
```

The last option ensures that the cell centers are only written out once in the subdirectory `casename/0`. There you find the four new files `C`, `Cx`, `Cy`, and `Cz`. In `C`, there are the positions of the cell centers as vectors. The other files only contain the coordinates of one particular direction. This step is only necessary if the grid is not uniform.

In our special case, we had two blocks with different cell heights. The block boundary has to be determined from the cell centers, as well as the two cell heights `dya` and `dyb`. Then the position of the surface, the film thickness δ, can be calculated by the summation of c_{VOF} multiplied by the cell height. From δ, we can find the two cell centers between which the surface is located. The surface temperature is calculated from

$$T_s = T_1 + (T_2 - T_1)\frac{\delta - y_1}{y_2 - y_1}. \tag{7.7}$$

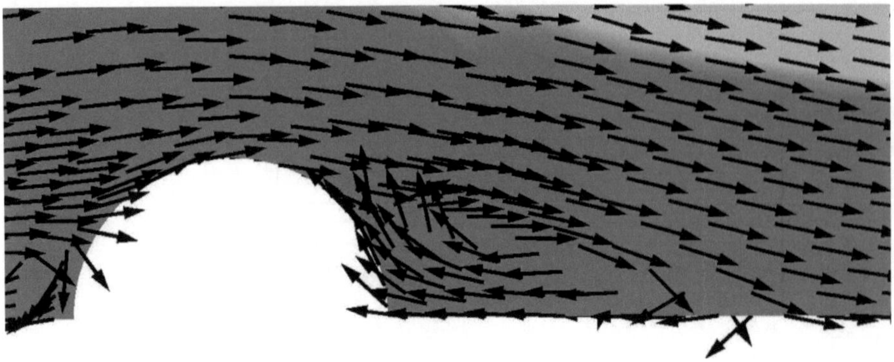

Figure 7.15: The wake of a wire as seen with OpenFOAM.

The index 1 (2) refers to the lower (higher) neighboring cell. This procedure is performed for $x > L_{\text{cut}}$ behind the wires where the numerical grid is rectangular.

Is a thermal simulation absolutely necessary in connection with turbulence wires? It is not mandatory. An idea would be to calculate the surface area and relate it to the interface area of a smooth file. Such a measure of waviness may also serve well to say which distance between the wires should be preferred.

Most technical details are given in the Appendices. *It should be apparent now that OpenFOAM is a powerful tool for the direct numerical simulation of falling liquid films.*

8 A Lesson from FS3D

Free Surface 3D (FS3D) is a finite volume based in-house code which employs the volume of fluid (VOF) method. It was originally developed by Rieber and Frohn [123]. At the Center of Smart Interfaces of the Technical University of Darmstadt, research on the simulation of wavy laminar falling films was performed with FS3D [11, 10]. In this chapter, an important lesson from these numerical experiments is outlined. The influence of surface tension models on the hydrodynamics was investigated. The three tested models were:

- continuum surface stress (CSS) model [83]
- balanced continuum surface force (bCSF) model
- unbalanced continuum surface force (uCSF) model

The CSF method was introduced in Section 5.5. What is called "balanced" and "unbalanced" is explained in a moment. All three surface tension models approximate the Laplace Force, equation (5.50). CSS as well as CSF define a volume force in a layer of the numerical grid where the gradient ∇c_{VOF} does not vanish. In the case of CSS, this force per volume is

$$\boldsymbol{F}_{\Sigma,\mathrm{CSS}} = \nabla \cdot \left[\sigma |\nabla c_{\mathrm{VOF}}| (\boldsymbol{I} - \boldsymbol{nn})\right], \tag{8.1}$$

where σ is the surface tension, c_{VOF} the volume of fluid scalar, $\boldsymbol{I}$ the identity matrix, and $\boldsymbol{n}$ the unit vector normal to the surface. Note that the sign of $\boldsymbol{n}$ does not matter here. In case of CSF, the force per volume is obtained from

$$\boldsymbol{F}_{\Sigma,\mathrm{CSF}} = \sigma \kappa \nabla c_{\mathrm{VOF}}. \tag{8.2}$$

The letter κ denotes the curvature of the interface. The force per volume $\boldsymbol{F}_\Sigma$ is integrated into the Navier–Stokes equations:

$$\rho \frac{\partial \boldsymbol{U}}{\partial t} + \rho(\boldsymbol{U} \cdot \nabla)\boldsymbol{U} + \nabla p = \mu \nabla^2 \boldsymbol{U} + \rho \boldsymbol{g} + \boldsymbol{F}_\Sigma \tag{8.3}$$

Note that $\boldsymbol{F}_\Sigma$ contains ∇c_{VOF}. In equation (8.3) appears also the gradient of another scalar, i. e., the pressure. In the case of bCSF, ∇c_{VOF} and the gradient of the pressure, ∇p, are calculated in the same manner, i. e., the discretization scheme is the same. Furthermore, the simple geometry of the mesh was of advantage to calculate the film thickness $\delta(x_i)$. For a certain streamwise coordinate x_i, all c_{VOF} of the cells perpendicular to the wall were summed up and multiplied with the uniform cell height Δy. The curvature was determined via the formula

$$\kappa = \frac{\frac{\partial^2 \delta}{\partial x^2}}{\left(1 + \left(\frac{\partial \delta}{\partial x}\right)^2\right)^{3/2}}. \tag{8.4}$$

https://doi.org/10.1515/9783110592337-008

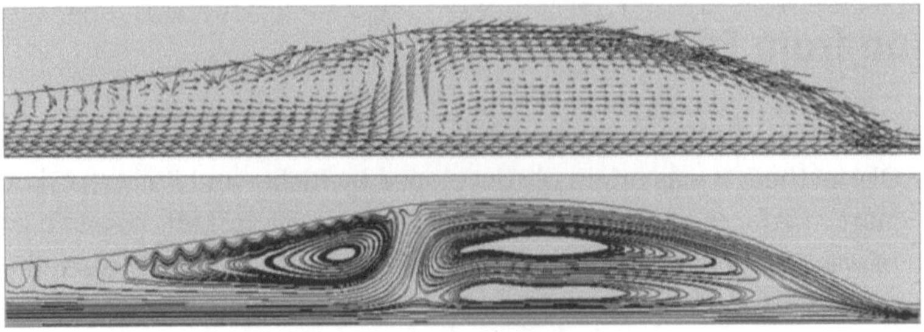

Figure 8.1: Vortex structure in the solitary wave, CSS.

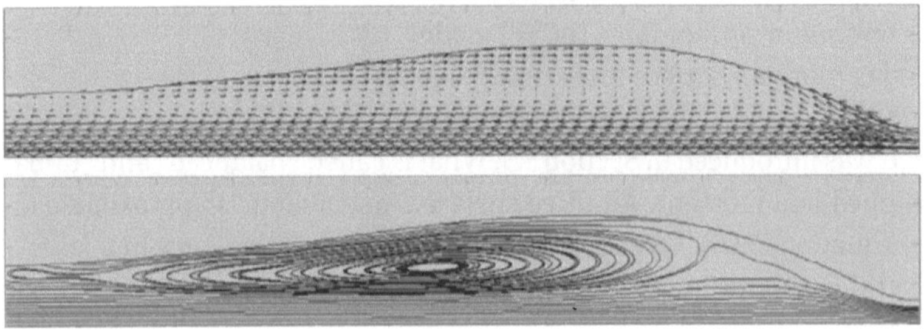

Figure 8.2: Vortex structure in the solitary wave, bCSF.

In the case of uCSF, no special care was taken for a compatible discretization of ∇c_{VOF} and ∇p. The surface curvature was obtained by differentiating the c_{VOF}-field.

With uCSF, the waves looked realistic at lower resolutions. However, when the grid was more and more refined, they looked unphysical. Therefore, this approach was dismissed.

Reasonable looking solutions were obtained from CSS and bCSF at all resolutions. However, when we went into the reference frame of the solitary wave, apparent differences were observed concerning the vortex structure. In Figure 8.1, there are three distinct vortices in the roll wave in case of the CSS model. On the other hand, with the bCSF model there is only one vortex, Figure 8.2. The question of the vortex structure has already been discussed in Section 7.3. The references [111, 112, 12] give experimental evidence for a single vortex. However, the CSS result was taken seriously and further tests were performed. They showed that bCSF can simulate a falling film reliably, whereas CSS has deficiencies if you take a closer look.

This summary of Albert et al. [11] is admittedly very concise. What is the lesson from these numerical experiments? *Although two formulas may be mathematically equivalent, after their numerical implementation, they are usually not the same. The results of a simulation have to be checked.*

9 Original programs

This chapter describes programs that can capture the whole saltwater film. In the case of the EasyMED desalination plant, the film ran down a hight of about 1.5 m. All along this chapter, no disturbance by turbulence wires is taken into account. From Section 9.1 up to Section 9.3, the surface of the falling film is assumed to be smooth. This is by no means an oversimplification, since important questions can now be addressed whose answers will remain valid as an acceptable approximation to the real situation. The potential of the long wave equations is tested in Section 9.4. As a refinement of the models with a smooth surface, harmonic waves are treated in Section 9.5. Furthermore, a kind of hybrid modeling with input from OpenFOAM will be examined in Section 9.6. Does the excitation at the inlet always has to be harmonic? Of course not. Random excitation will also be treated in that section. Finally, the evaporation rates of pure water and saltwater are compared, Section 9.7. In this chapter, the mass transport of the salt is given consideration, whereas this aspect has been neglected so far. Many original programs will be outlined here. They all have been written in C++. It is better to repeat certain formulae than to refer to their first appearance. It has the advantage that the reader does not have to look up so many equations.

9.1 One-dimensional model with VOF

Equation (4.25) in Nusselt's theory has been linearized by calculating the first two terms of the Taylor expansion at $x = 0$ so that, according to our numerical algorithm, the film thickness δ a distance x downward the inlet can be obtained by

$$\delta(x) = \delta_{\text{in}} - \int_0^x \frac{\lambda_L \nu_L (\partial T/\partial y)_i}{\delta^2(\tilde{x}) \cdot \rho_L \cdot g \cdot \Delta h} d\tilde{x}, \tag{9.1}$$

where δ_{in} denotes the film thickness at the inlet, λ the thermal conductivity, and Δh the evaporation enthalpy. The index "L" refers to the liquid. Notice that the properties of the liquid can now be regarded as a function of x. The temperature profile in the film is determined from a 1D heat diffusion equation:

$$\frac{\partial T}{\partial t} = D_{\text{th}} \frac{\partial^2 T}{\partial y^2} \tag{9.2}$$

Equation (9.2) can be solved analytically for short times in terms of the error function erf, and for long times by using Fourier's series. However, for our purposes it was most efficient to solve it numerically. The time step Δt is connected with Δx by

$$\Delta t = \frac{\Delta x}{u_I} = \frac{2\nu_L \cdot \Delta x}{g \cdot \delta^2(x)}. \tag{9.3}$$

https://doi.org/10.1515/9783110592337-009

The velocity u_I of the film interface has been calculated from the Nusselt profile, equation (4.15). It is reasonable to proceed in time with the velocity at the interface since the evaporation takes place at the surface of the film.

The 1D model outlined gives the same temperature profile as a 2D model which also accounts for convection after a relatively short entrance length.

9.1.1 Effect of salinity

One has to consider the evaporation of the solvent (water) resulting in a decrease of the film thickness to obtain the salt concentration inside the evaporating film, thus leading to a higher salinity. The salt is mainly transported in the film by diffusion, thus for approximately constant solution density,

$$\frac{\partial S}{\partial t} = D_s \frac{\partial^2 S}{\partial y^2}, \tag{9.4}$$

where S is the water salinity and D_s the salt diffusion coefficient.

Furthermore, it must be imposed that the amount of salt inside the film remains constant, for only the water evaporates. So, the following iterative algorithm has been implemented:

1. Uniformly set initial salinity, $S_{\text{in}} = 35\,\text{g/kg}$.
2. Proceed in time: $\Delta t = \Delta x/u_I = 2\nu_L \cdot \Delta x/(g \cdot \delta_0^2)$, where δ_0 is the film thickness of the previous iteration.
3. Reduce the film thickness according to $\Delta\delta = (\lambda_L \nu_L (\partial T/\partial y)_i \cdot \Delta x)/(\delta^2(x) \cdot \rho_L \cdot g \cdot \Delta h)$. From this reduction, deduce the change of the volume of fluid scalar, Δc_{VOF}.
4. At interfacial and neighboring cell set salinity equal to $S = S_0 + S_{\text{in}} \cdot \Delta c_{\text{VOF}}$. ($S_0$ is the salinity of the previous iteration.)
5. Solve the diffusion equation (9.4).
6. Normalize S by imposing $\int_0^\delta S\,dy = \text{const.}$
7. Set δ_0 equal to the new film thickness δ and S_0 equal to the new salinity S. Restart at step 2.

Step 4 is supposed to insure that when all the water in a cell at the interface has evaporated, the amount of the salt has been transferred to the neighboring liquid cell inside the film. This basic idea of the algorithm is illustrated in Figure 9.1.

9.1.2 Boiling point elevation

The salt concentration at the interface affects the temperature at which the water can evaporate. The higher the concentration the higher is the evaporation temperature

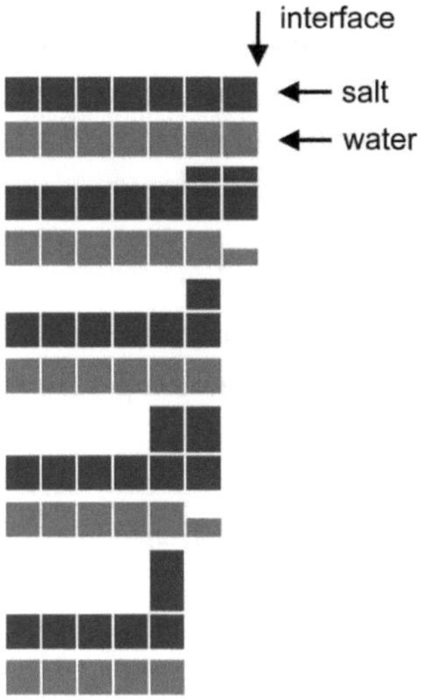

Figure 9.1: Increase of salinity at the interface due to the evaporation.

while the pressure remains constant. This phenomenon is called "boiling point elevation." A separate function for its calculation is employed in the program. How the boiling point elevation is updated in each time step, will be detailed in Section 9.1.3.

9.1.3 Outline of the program

The program shall simulate a laminar, evaporating seawater film in one dimension (1D). The film runs down a vertical, heated plate with constant wall temperature T_{wall}. The surface temperature is denoted by $T_{surface}$. The resulting film thickness is compared with Nusselt's waterskin theory, equation (4.25). To keep the amount of salt inside the film constant, a normalization is performed. The explicit scheme for the progress in time is employed. Central differencing is used for the spatial derivatives. The number of cells as well as the cell size remain constant. The volume of fluid scalar c_{VOF} yields the position of the interface.

At the beginning, the Reynolds' number (alternatively the flow rate and the plate width) and the temperature of the wall have to be known. Also the temperature of the water at the inlet has to be chosen. Additionally, a value for the seawater salinity, e. g., 35 g/kg, must be given. Equation (4.19) yields the film thickness, equation (4.16) the velocity at the surface. As the simulation is one-dimensional, only the number of cells `Ny` in the *y* direction are important for the grid. The new and the old temperatures are defined in an array. (In C++, there is also the possibility to employ the data type "vector.") In C and C++, the index runs from 0 to `Ny-1`. Such an index shall be denoted by `i`. In Chapter 5, the finite volume method has been introduced. We are going to apply it to our case. Instead of the indices "W," "P," and "E," we will use `i-1`, `i`, and `i+1`, respectively. The cell next to the wall has the index 0. Initially, the surface is the eastern boundary of the cell with the index `Ny-1`. The discretization of equation (9.2) is

$$\frac{T_i^{\text{new}} - T_i}{\Delta t} = D_{\text{th}} \frac{T_{i-1} - 2T_i + T_{i+1}}{(\Delta y)^2}. \tag{9.5}$$

The superscript "new" refers to the new time step. It is implied that all the other temperatures belong to the old time step. Equation (9.5) is solved for the new temperature:

$$T_i^{\text{new}} = \frac{D_{\text{th}}\Delta t}{(\Delta y)^2}(T_{i-1} + T_{i+1}) + \left(1 - 2\frac{D_{\text{th}}\Delta t}{(\Delta y)^2}\right)T_i \tag{9.6}$$

This can be expressed more neatly:

$$\begin{aligned} T_i^{\text{new}} &= a_W T_{i-1} + a_P T_i + a_E T_{i+1} \\ \text{with} \quad a_W &= a_E = \frac{D_{\text{th}}\Delta t}{(\Delta y)^2} \quad \text{and} \quad a_P = 1 - a_W - a_E \end{aligned} \tag{9.7}$$

How is a certain temperature imposed at the boundary? To see this, we have a look at the cell next to the wall,

$$\frac{T_0^{\text{new}} - T_0}{\Delta t} = \frac{D_{\text{th}}}{\Delta y}\left(\frac{(T_1 - T_0)}{\Delta y} - \frac{(T_0 - T_{\text{wall}})}{0.5\Delta y}\right). \tag{9.8}$$

The western boundary of the cell has the temperature T_{wall}. It is only $0.5\Delta y$ away from the cell center. Again, we solve for the new temperature:

$$T_0^{\text{new}} = \frac{D_{\text{th}}\Delta t}{(\Delta y)^2}T_1 + \left(1 - 3\frac{D_{\text{th}}\Delta t}{(\Delta y)^2}\right)T_0 + 2\frac{D_{\text{th}}\Delta t}{(\Delta y)^2}T_{\text{wall}} \tag{9.9}$$

This can be rewritten as

$$\begin{aligned} T_0^{\text{new}} &= a_P T_0 + a_E T_1 + R_u, \\ \text{where} \quad a_E &= \frac{D_{\text{th}}\Delta t}{(\Delta y)^2}, \quad a_P = 1 - 3a_E \quad \text{and} \quad R_u = 2a_E T_{\text{wall}}. \end{aligned} \tag{9.10}$$

At the eastern boundary of the last cell, we impose the surface temperature T_{surface}. The procedure is analogous to the wall temperature. It yields

$$\begin{aligned} T_{\text{Ny}-1}^{\text{new}} &= a_W T_{\text{Ny}-2} + a_P T_{\text{Ny}-1} + R_u, \\ \text{where} \quad a_W &= \frac{D_{\text{th}}\Delta t}{(\Delta y)^2}, \quad a_P = 1 - 3a_W \quad \text{and} \quad R_u = 2a_W T_{\text{surface}}. \end{aligned} \tag{9.11}$$

The salt diffusion equation, equation (9.4), is discretized in the same way:

$$\frac{S_i^{\text{new}} - S_i}{\Delta t} = D_s \frac{S_{i-1} - 2S_i + S_{i+1}}{(\Delta y)^2} \tag{9.12}$$

Thus the new salinity is calculated from

$$S_i^{\text{new}} = a_W S_{i-1} + a_P S_i + a_E S_{i+1}$$
$$\text{with} \quad a_W = a_E = \frac{D_s \Delta t}{(\Delta y)^2} \quad \text{and} \quad a_P = 1 - a_W - a_E. \tag{9.13}$$

The boundaries need a special treatment. There is no salt diffusion through the wall which yields a zero gradient boundary condition:

$$\left.\frac{\partial S}{\partial y}\right|_{\text{wall}} = 0 \tag{9.14}$$

We obtain for the first cell

$$S_0^{\text{new}} = a_P S_0 + a_E S_1,$$
$$\text{where} \quad a_E = \frac{D_s \Delta t}{(\Delta y)^2} \quad \text{and} \quad a_P = 1 - a_E. \tag{9.15}$$

Now that the discretization of the two diffusion equations is clear, we can turn to the easier aspects of the program. Equation (9.3) yields the initial time step Δt_{ini}. The number of time steps, N_t, is determined at the beginning:

$$N_t = \frac{L}{u_{\max} \Delta t_{\text{ini}}}, \tag{9.16}$$

where L is the length of the heating plate. Note that the floating point number on the right-hand side has to be transformed into an integer. According to equation (9.3), the time step size depends on δ^2. After each time step, it is updated:

$$\Delta t^{\text{new}} = \Delta t^{\text{old}} \left(\frac{\delta^{\text{old}}}{\delta^{\text{new}}} \right)^2 \tag{9.17}$$

The new film thickness is calculated from the sum of all c_{VOF} multiplied by Δy. The reduction of c_{VOF} follows from $\Delta\delta$:

$$\Delta c_{\text{VOF}} = \frac{\Delta\delta}{\Delta y} = -\lambda \left.\frac{\partial T}{\partial y}\right|_{\text{surface}} \frac{\Delta t}{2\rho \Delta h \, \Delta y} \tag{9.18}$$

According to this equation, Δc_{VOF} can be larger than 1. The cell with the interface is marked with the integer Nsurf. Initially, Nsurf = Ny. From Δc_{VOF}, the salinity at the surface is obtained:

$$S_{\text{Nsurf}-1}^{\text{new}} = S_{\text{Nsurf}-1}^{\text{old}} + S_{\text{in}} \Delta c_{\text{VOF}}$$
$$S_{\text{Nsurf}-2}^{\text{new}} = S_{\text{Nsurf}-1}^{\text{new}} \tag{9.19}$$

Calculating the elevated saturation temperature for each time step is not as trivial as it may seem. The function bpe yields the boiling point elevation at a given temperature on account of the salinity. There are two variables, `eto` and `etn`, elevated temperature old and new, respectively. Before entering the time loop, we initialize `eto`. Later we determine `etn` for each time step:

$$\begin{aligned} t = 0: &\qquad \text{eto} = \text{bpe}(T_{\text{in}}, S_{\text{in}}) \\ t > 0: &\qquad \text{etn} = \text{bpe}(T_{\text{surface}}, S_{\text{Nsurf}-1}) \\ &\qquad T_{\text{surface}} = T_{\text{surface}} + \text{etn} - \text{eto} \\ &\qquad \text{eto} = \text{etn} \end{aligned} \tag{9.20}$$

(The third equation only makes sense in a computer program.) With the aid of the surface temperature, the gradient is calculated:

$$\left.\frac{\partial T}{\partial y}\right|_{\text{surface}} = \frac{T_{\text{surface}} - T_{\text{Nsurf}-1}}{0.5\Delta y} \tag{9.21}$$

As the grid is one-dimensional, we can choose a large number of cells in the y direction, e. g., Ny = 100. Therefore, the error is small if we do not reconstruct the exact position of the surface inside a cell. Instead, the eastern boundary of the surface cell always has the temperature T_{surface}. Since the procedure for the higher salinity at the surface is not conservative, we need to normalize S. This is done in a way that guarantees that S does not become smaller than S_{in}:

$$\begin{aligned} t = 0: &\quad \text{fnorm} = \delta S_{\text{in}} \\ t > 0: &\quad \text{fint} = \sum_{i=0}^{\text{Nsurf}-1} \Delta y c_{\text{VOF},i} S_i \\ &\quad S_i = S_{\text{in}} + (S_i - S_{\text{in}})\text{fnorm}/\text{fint} \end{aligned} \tag{9.22}$$

What if the salt diffusivity D_s is not known? A workaround may be

$$D_s \approx \frac{D_{\text{th}}}{100} \quad \text{(workaround!).} \tag{9.23}$$

D_s often had a value of $1.6 \cdot 10^{-9}\ \text{m}^2/\text{s}$.

Just to avoid confusion, here is a chronological list of what happens in the time loop:

1. Is the time step size small enough for the explicit scheme?
2. Are there more than four liquid cells left?
3. New film thickness from c_{VOF} and Δy.
4. Adjust film thickness, equation (9.17).
5. T_i^{new} calculated, equations (9.7), (9.10), (9.11). In equation (9.11), Nsurf is used instead of Ny.
6. With the help of the function `bpe`, the boiling point elevation is taken into account. New surface temperature, equation (9.20).

7. Temperature gradient at surface determined, equation (9.21).
8. Reduction of c_{VOF}, equation (9.18).
9. Salt diffusion inside the film, equations (9.13), (9.15).
10. New salinity at surface, equation (9.19).
11. Normalize salinity, equation (9.22).

9.1.4 Results of 1D simulations

The results obtained by the 1D model are summarized in Table 9.1. The water temperature at the inlet was $T_{\mathrm{in}} = 60\,°\mathrm{C}$ in these numerical experiments. The temperature of the plate was $T_w = 65\,°\mathrm{C}$. The flow rate was varied as parameter giving the Reynolds' number Re, the film thickness δ_{in}, and the maximum velocity u_I at the inlet as listed in Table 9.1. The film thickness δ_{end}, the water salinity at the surface S_{endi} and the interface temperature T_{endi} at the bottom of the plate are the simulation results. The seawater properties were determined in the calculations according to the formulas of EasyMED WP 1 Deliverable D2 Annex 3. A public source of seawater properties is [128]. As can be deduced from Table 9.1, the effect of evaporation is rather small for a flow rate of 150 l/h, but becomes stronger as the flow rate decreases.

Table 9.1: Results of 1D model calculations for various flow rates, $T_{\mathrm{in}} = 60\,°\mathrm{C}$, $T_w = 65\,°\mathrm{C}$, $S_{\mathrm{in}} = 35\,\mathrm{g/kg}$.

Flow rate l/h	Re	δ_{in} mm	u_I m/s	δ_{end} mm	$\Delta\delta/\delta_{\mathrm{in}}$ %	S_{endi} g/kg	T_{endi} °C
150	220	0.25	0.63	0.244	2.5	44.8	60.3
100	147	0.21	0.48	0.201	4.8	50.3	60.5
50	73	0.17	0.30	0.145	12.9	65.0	61.0
25	37	0.13	0.19	0.074	44.7	102.2	62.9

Figures 9.2 and 9.3 display the salinity profiles in the film at some distances x from the inlet for flow rates listed in Table 9.1. The position of the interface at various x is also indicated there. As follows from Figure 9.2 (top), at a flow rate of 150 l/h, the film thickness is almost constant. In contrast, it is seen from Figure 9.3 (bottom) that for a flow rate of 25 l/h about half of the film evaporates thus leading to a high salt concentration.

According to the 1D model for a completely homogeneous, laminar film with a flow rate of 100 l/h, the evaporation rate is 2.3 %. The measured value is 4 % for film flow with wires. Considering the fact that the plate in the pilot plant is not isothermal as assumed in our calculation, the discrepancy of the two values seems acceptable.

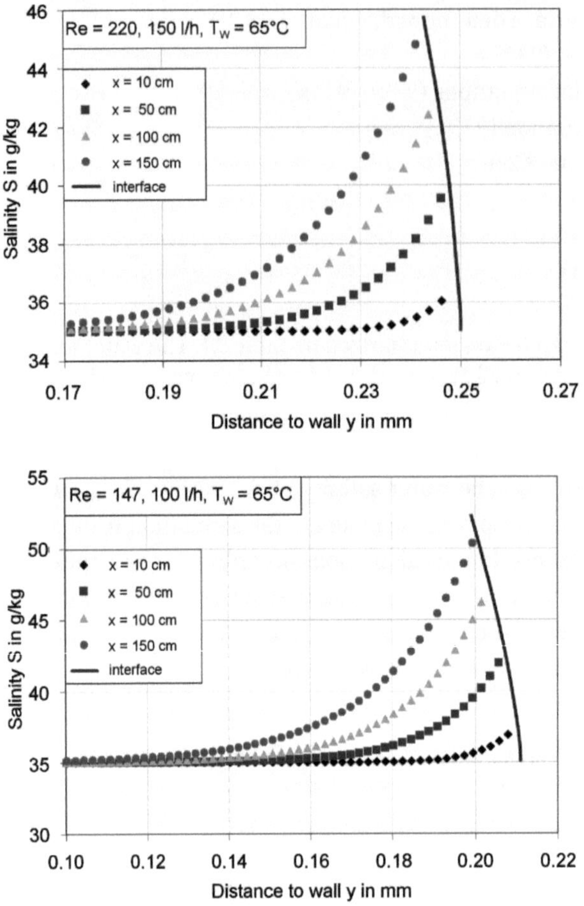

Figure 9.2: Salinity distribution inside evaporating seawater film and position of the interface for a flow rate of 150 l/h (top) and 100 l/h (bottom). This kind of presentation was earlier employed by Unterberg and Edwards [141].

(Furthermore, the plate in the pilot plant is not absolutely plane, there is the 3D influence of the wires, the shear stresses of the vapor, and the plate is not completely covered with liquid. All these items are not considered in the 1D model.)

9.1.5 Refinement of 1D model

The 1D model with fixed numerical grid was later compared to a 2D model with adaptive grid. Both programs were supposed to simulate an evaporating seawater film that is heated by a plate of constant temperature. Before the 2D model is described in the next section, a major change also applied in the 1D program shall be noted here. As

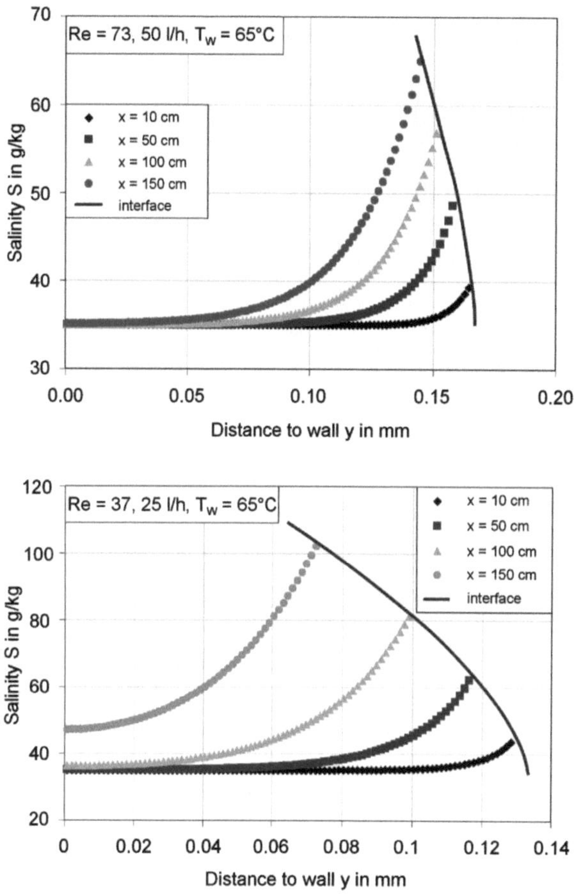

Figure 9.3: Salinity distribution inside evaporating seawater film and position of the interface for a flow rate of 50 l/h (top) and 25 l/h (bottom). This kind of presentation was earlier employed by Unterberg and Edwards [141].

the film is heated, its properties will change, namely the density ρ and kinematic viscosity ν. Assuming a Nusselt velocity profile, equation (4.15), the mass flow rate is (cf. equation (4.18))

$$\dot{M} = \frac{\rho}{3\nu} \cdot g \cdot Z \cdot \delta^3, \tag{9.24}$$

where g is the gravitational acceleration, Z the width of the film, and δ the film thickness. As follows from equation (4.25), the change of the film thickness δ along the flow direction x depends on the physical properties and the driving temperature difference $T_w - T_{sat}$. Neglecting for the moment the evaporation effect, that is to say, omitting the second term in the big brackets of equation (4.25), equation (9.24) gives

$$\delta = \left(\frac{\rho_{\text{in}}}{\rho} \frac{\nu}{\nu_{\text{in}}} \right)^{1/3} \delta_{\text{in}}, \tag{9.25}$$

where ν and ρ are the average properties in the cross-sectional area of the film at the distance x. Equation (9.25) refers to the values at the inlet. If we substitute the subscript “in” with “0,” we get

$$\delta = \left(\frac{\rho_0}{\rho} \frac{\nu}{\nu_0} \right)^{1/3} \delta_0, \tag{9.26}$$

i. e., the values of the previous iteration are employed as reference. The major change is a readjustment of the film thickness according to equation (9.26) accounting for the variation of the liquid properties.

9.2 Two-dimensional simulations with adaptive grid

In this section, programs shall be introduced which employed adaptive numerical grids. The term “adaptive” is supposed to mean that the number of cells filled with liquid was constant whereas the cell size changed on account of the evaporation.

It was observed in earlier investigations that the Nusselt velocity profile adequately describes the velocities in a hydrodynamically developed, undisturbed film. With the velocity given in equation (4.15), the problem simplifies considerably. A computer program has been written that solves for the two-dimensional (2D) time-marching convection-diffusion temperature equation:

$$\frac{\partial T}{\partial t} + u \frac{\partial T}{\partial x} = D_{\text{th}} \left(\frac{\partial^2 T}{\partial x^2} + \frac{\partial^2 T}{\partial y^2} \right) \tag{9.27}$$

In equation (9.27), T is the temperature, t the time, x and y the distances from the inlet and the wall, respectively, and D_{th} the thermal diffusivity of the liquid which is assumed to be constant. The temperature of the heating wall was kept at $T_w = 65\,°\text{C}$. The inlet temperature of the water was five degrees below the wall temperature. This temperature was taken as the boiling point ($T_{\text{in}} = T_{\text{sat}} = T_i$) which was also the temperature at the vapor-liquid interface in the case of evaporation.

The question addressed in the simulation was whether there is a developed temperature profile in the film and if so, at what distance from the inlet it is established. The answer was that after less than approximately 10 cm downstream of the inlet a linear temperature profile is reached (cf. Figure 9.4) satisfying

$$T = T_w - (T_w - T_{\text{sat}}) \frac{y}{\delta}. \tag{9.28}$$

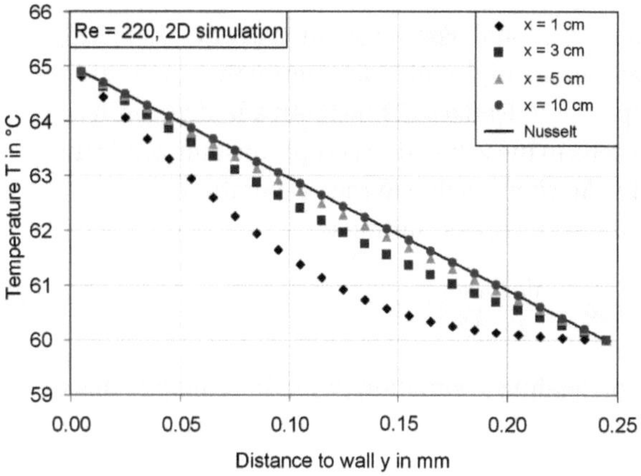

Figure 9.4: Temperature distribution in the entrance region, $\delta_{in} = 0.25$ mm.

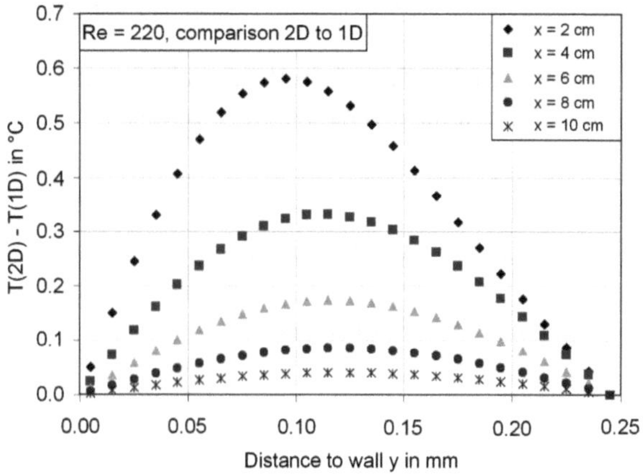

Figure 9.5: Difference of film temperature between 1D and 2D models, $\delta_{in} = 0.25$ mm.

The next important issue treated was the difference of the temperature profiles obtained with the 2D and 1D model. The result is displayed in Figure 9.5. At the boundaries $y = 0$ and $y = \delta$, the temperatures are kept at the same values in both models. It is clearly seen that the deviation of the models decreases more and more downstream and becomes negligible for $y > 10$ cm. This brings us closer to the aim of simulating the whole film one-dimensionally, for the 1D model is computationally much more efficient. The error associated with this simplification is negligibly small.

The governing equation for the mass transport in our two-dimensional case is

$$\frac{\partial S}{\partial t} + u\frac{\partial S}{\partial x} = D_s\left(\frac{\partial^2 S}{\partial x^2} + \frac{\partial^2 S}{\partial y^2}\right) \tag{9.29}$$

The velocity is supposed to be unidirectional. For the solution of equation (9.29), the finite volume method is employed with a fully implicit scheme for the temporal change and central differences for the spatial derivatives [50, 143]. An adaptive mesh is used which means that the number of cells in the y direction is kept constant, while the film thickness δ, and thus the cell size Δy changes due to the evaporation. During a time step Δt, the film thickness δ is reduced by $\Delta_e\delta$ according to

$$\Delta_e\delta = \frac{\lambda|\partial T/\partial y|_i}{2\rho \cdot \Delta h} \cdot \Delta t. \tag{9.30}$$

(See Annex A for a neat calculation leading to equation (9.30). In principle, this result has already been obtained in Section 9.1.) Also the convective change must be taken into account, thus

$$\Delta\delta = -u(\partial\delta/\partial x) \cdot \Delta t - \Delta_e\delta. \tag{9.31}$$

The velocity is calculated according to the Nusselt formula, equation (4.15), depending on the fluid properties and the film thickness.

Water evaporates on the film surface leaving the salt in the remaining seawater film. This leads to an increase of the salinity in the cell next to the interface:

$$\Delta S = S_{\text{in}} \cdot (\Delta_e\delta/\Delta y_{\text{in}}). \tag{9.32}$$

In equation (9.32), S_{in} denotes the initial salinity, $\Delta_e\delta$ the reduction of the film thickness due to evaporation, equation (9.30), and Δy_{in} the initial cell size in the y direction. Since the amount of salt in the film does not change, the conservation integral

$$\int_0^{\delta} S\,dy = \text{constant} \tag{9.33}$$

is required to hold.

The most difficult task is to solve for T and S on the two-dimensional grid. A description of the algorithm would take very long. Eventually, it will be shown in Section 9.2.1 that one-dimensional simulations are sufficient for our purposes. Therefore, the reader is advised to download programs by Peric [1] to see how such solvers are programmed efficiently.

9.2.1 Comparison of 1D and 2D models

A comparison was performed between the 1D and 2D model that both describe a falling evaporating seawater film heated by a vertical plate. The wall temperature was kept constant at $T_w = 65\,°\text{C}$ and the seawater inlet temperature and salinity were $T_{\text{in}} = 60\,°\text{C}$ and $S_{\text{in}} = 35\,\text{g/kg}$, respectively.

Table 9.2: Comparison between the 1D and 2D models, $S_{in} = 35$ g/kg.

Flow rate l/h	Re	$\Delta\delta_{end}/\delta_{in}$ %		S_{endi} g/kg		T_{endi} °C	
		1D	2D	1D	2D	1D	2D
25	37	53.7	60.3	92.44	108.4	61.55	62.11
50	73	13.3	12.5	52.66	53.76	60.17	60.14
100	147	5.3	4.4	43.62	44.09	60.06	60.05
150	220	3.5	2.5	41.77	41.19	60.04	60.03

Table 9.2 shows a comparison of the models for selected Reynolds' numbers. The flow rates corresponding to Re in the case of the EasyMED desalination plant are also displayed. The first comparison deals with the decrease of the film thickness at the bottom of the 1.5 m long plate. The reduction of the film thickness is measured in terms of the initial film thickness. The second and third comparison criteria are the salinity at the interface S_{endi} and the temperature of the interface T_{endi} both determined at the bottom of the plate, the index "i" referring to the interface. As the interface salinity S_{endi} depends on the number of cells in the y direction, this number was the same in the 1D and 2D program. The interface temperature T_{endi} is higher than 60 °C since the average salinity in the film increases due to the evaporation. It was observed in the numerical experiments that after a short entrance region of approximately 10 cm the temperature profiles of the 1D and the 2D program were both linear, equation (9.28).

It can be seen from Table 9.2 that the agreement of the 1D and 2D models is good. The 1D program should be preferred considering that calculations with the 2D program took several days, whereas the results of the 1D model were obtained in some minutes.

9.3 Conjugate heat transfer

For the reasons stated above, we adopted the 1D program for simulating the conjugate heat transfer between the condensing water film on one side of the plate and the evaporating seawater film on the other side; cf. Figure 9.6. In this program, an adaptive grid was used, i. e., the number of cells was constant whereas their coordinates varied according to the changing film thicknesses. The film thickness of the evaporating saltwater film is calculated by

$$\delta(x) = \left(\delta_0^4 - \int_0^x \frac{4\lambda_L \nu_L (T_w - T_{sat})}{\rho_L g \Delta h} \cdot d\tilde{x} \right)^{1/4} . \tag{9.34}$$

Note that the above relation is more accurate than equation (4.25) of the Nusselt theory, since all quantities in the integral kernel can be treated as functions of the distance from the inlet x. The same equation is employed for the condensing water film.

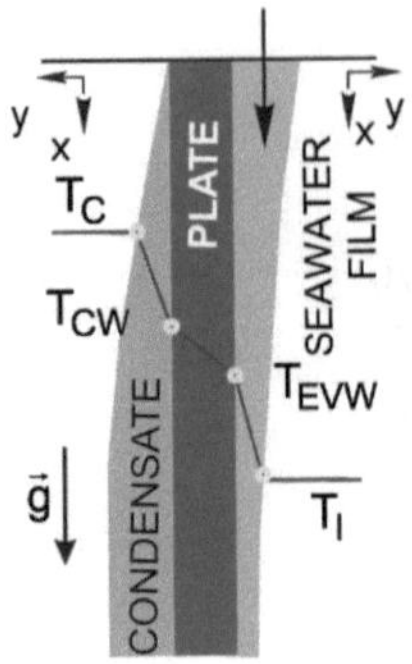

Figure 9.6: Conjugate heat transfer between condenser and evaporator.

In this case the variables in equation (9.34) may have other values, for the temperature on the other side of the plate and the initial film thickness are different and the condensate does not contain any salt.

To calculate the salinity in the seawater film, again equations (9.32) and (9.33) are employed and a 1D diffusion equation is solved in the domain of the saltwater film. The energy equation is simultaneously solved for in the condensing film, in the plate, and in the saltwater film. The surface temperature of the seawater film is the saturation temperature which basically may change along the film flow because of the salinity increase due to the evaporation. The simulation is transient. The elapsed time is worked out in the distance from the inlet x according to the velocity at the interface of the evaporating film.

Numerical experiments were performed with a temperature of 65 °C for the heating, condensing vapor, and an initial temperature of 60 °C for the seawater film. Again, a temperature difference of 5 °C is assumed. However, the simulation is more realistic now, since the temperature drop in the condensate film and inside the plate is determined and the wall temperatures are not constant any more. The initial salinity was 35 g/kg. An initial thickness of the condensate film of 0.05 mm was chosen. The stainless steel plate was 1 mm thick.

As an example, Figure 9.7 shows the temperature distributions along the falling film at a Reynolds' number of Re = 73. The temperature of the vapor (T_C), the temperature of the wall on the condensation side (T_{CW}) and on the evaporation side (T_{EVW}), and the interface temperature of the saltwater film (T_I) are displayed. The initial wall temperature is unknown. Two extreme initial conditions are displayed for that reason in Figure 9.7. It is seen from this figure that the initial wall temperature has a strong influence in the beginning, whereas in the long run the discrepancies disappear.

Table 9.3 summarizes the results. It shows the relative decrease of the seawater film thickness, the surface temperature of the evaporating film, the wall temperatures and the heat flux. The values listed in Table 9.3 all refer to the bottom of the plate which has a wetted height of 1.5 m. A comparison to the data in Table 9.2 reveals that

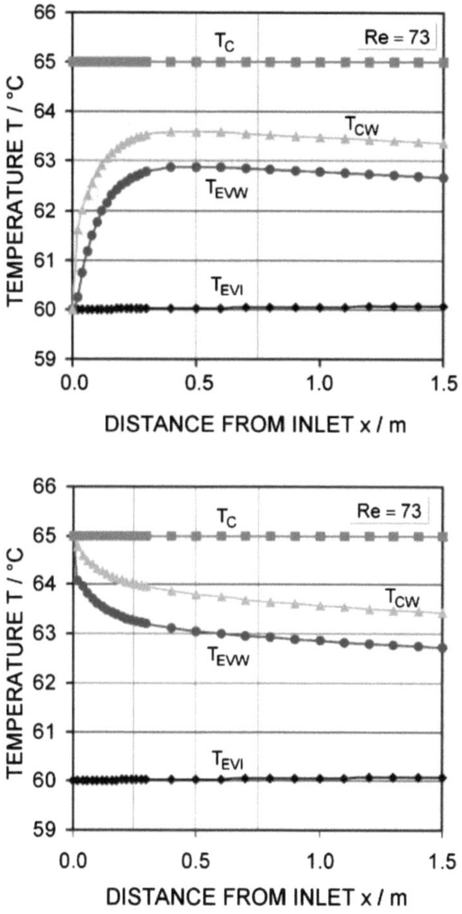

Figure 9.7: Temperatures at the boundaries of the two films in dependence on the distance from the inlet for two different initial wall temperatures.

Table 9.3: Results from 1D simulation of conjugate heat transfer.

Evaporating film				**Wall temperatures**		**Heat flux**
Flow rate l/h	Re	$\Delta\delta_{END}/\delta_{IN}$ %	T_{ENDI} °C	T_{EV} °C	T_{CON} °C	q_{END} kW/m^2
25	37	16.5	60.2	62.31	63.09	11.5
50	73	6.7	60.07	62.68	63.36	10
100	147	3	60.03	63	63.59	8.6
150	220	2	60.02	63.14	63.68	8

the evaporation is weaker now. This is due to a lower wall temperature, which arises from the temperature drop in the condensate film and inside the plate.

9.4 Long wave equations

The long wave equations are two coupled equations for the film thickness δ and the specific flow rate q_u. They have been introduced in Section 4.11. Equations (4.79) have been discretized with the finite difference method. Some details of the program and results will be presented here. To say it right away, it is not a success story. However, "The imperfections of a wise man make your rule, not the perfections of a fool!" This is an old wisdom cited in a physics book. Maybe the imperfections are not so wise. Anyway, the reader can learn from them.

Before some results are shown and discussed, we need to take a look on the numerical details of a program. Initially, the film is uniformly smooth. The thickness and specific flow rate are calculated from the Nusselt solution. The temporal development of the flow rate is discretized in the following way:

$$\begin{aligned}\frac{q_{u,i}^{\text{new}} - q_{u,i}}{\Delta t} = &-\frac{12}{5}\frac{q_{u,i}}{\delta_i}\frac{(q_{u,i+1} - q_{u,i-1})}{2\Delta x} + \frac{6}{5}\left(\frac{q_{u,i}}{\delta_i}\right)^2\frac{(\delta_{i+1} - \delta_{i-1})}{2\Delta x}\\ &- 3\nu_L\frac{q_{u,i}}{\delta_i^2} + g\delta_i\\ &+ \frac{\sigma}{\rho_L}\delta_i\frac{(\delta_{i+2} - 2\delta_{i+1} + 2\delta_{i-1} - \delta_{i-2})}{2(\Delta x)^3}\end{aligned} \tag{9.35}$$

The explicit temporal discretization is employed. The superscript "new" marks the value for the next time step. A superscript "old" is omitted, for it is easily understood that all the other values belong to the "old time step." For the spatial derivatives, the central difference scheme is used, according to the formula for some function f

$$\frac{\partial^n f}{\partial x^n} \approx \frac{\Delta^n f}{\Delta x^n} = \begin{cases} \frac{f_{i+1}^{(n-1)} - f_{i-1}^{(n-1)}}{2\Delta x} & : \quad n \text{ odd}\\ \frac{f_{i+1}^{(n-2)} - 2f_i^{(n-2)} + f_{i-1}^{(n-2)}}{\Delta x^2} & : \quad n \text{ even}\end{cases} \tag{9.36}$$

The second equation of equations (4.79) is treated in the same manner:

$$\frac{\delta_i^{\text{new}} - \delta_i}{\Delta t} = -\frac{(q_{u,i+1} - q_{u,i-1})}{2\Delta x} \tag{9.37}$$

At the inlet, the specific flow rate is modulated harmonically:

$$q_u(x, t) = q_{u,\text{Nusselt}}(1 + \varepsilon\sin(\omega t - k_x x)) \tag{9.38}$$

where ω is the frequency multiplied by 2π. The phase velocity of a harmonic wave yields k_x:

$$u_{\text{ph}} = \frac{\rho_L}{\mu_L}g\delta_0^2 = \frac{\omega}{k_x} \tag{9.39}$$

In equation (9.38), ε is usually set to 0.03. At the inlet, $x = 0$. The x dependency of equation (9.38) becomes import when q_u is to be calculated for $x < 0$. This is of interest for the first two cells in connection with the central difference scheme. From equation (9.38), the film thickness at the inlet follows

$$\delta(0,t) = \left(\frac{3\nu_L q_u(0,t)}{g}\right)^{1/3} \tag{9.40}$$

Although the perturbation at the inlet could not have reached the outlet, the film in its vicinity was very unstable. Therefore, a cut-off was calculated. It marks the number of cells to be processed.

$$N_{\text{xlimit}} = \min(\max((f_{\text{safety}} u_{\text{ph}} t/\Delta x), N_x/10), N_x), \tag{9.41}$$

where N_x is the total number of cells and u_{ph} the phase velocity of a harmonic wave; see equation (9.39). (In a computer program, the implementation of equation (9.41) may be a bit more complicated, since Nx and Nxlimit are integers, whereas the other variables are floating point numbers.) The safety factor f_{safety} is usually chosen between 1 and 1.5.

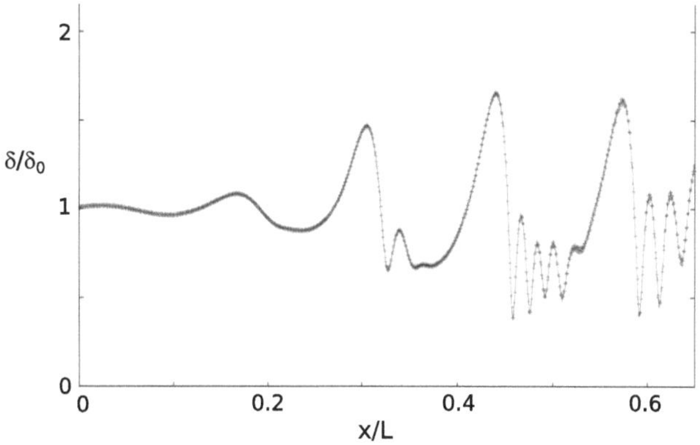

Figure 9.8: Solitary waves with the long wave equations at Re = 40.

Figure 9.8 shows the result of a successful simulation. The Reynolds' number was 40, water at 20 °C. The Weber number was 10.4. The initial film thickness was δ_0 = 0.2314 mm, which is also the thickness of the corresponding smooth film. The length of the control volume was $L = 600\delta_0 \approx 140$ mm. About three million time steps were calculated with a step size of $\Delta t = 8.868 \cdot 10^{-8}$ s. The length L was divided into 1500 cells with a size of $\Delta x = 0.0926$ mm. The safety factor for N_{xlimit} was 1.35. The excitation was performed with a frequency of 20 Hz and an amplitude of 0.03. Figure 9.8 nicely displays solitary waves. However, the thickness of the capillary waves is a bit too large. According to Kalliadasis et al. [72], this is a typical phenomenon if this type of equation is employed.

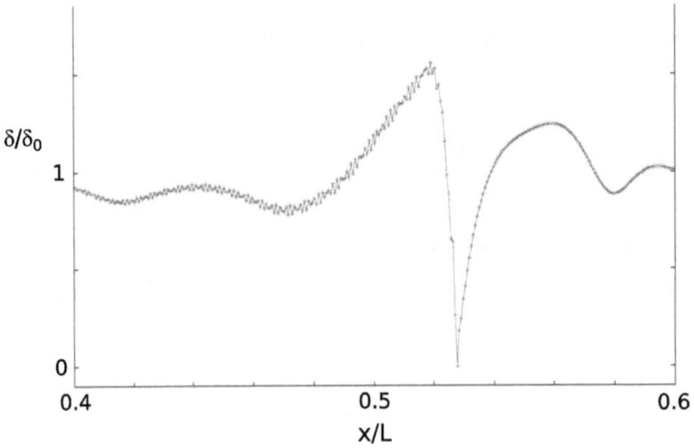

Figure 9.9: Film rupture at Re = 80.

Figure 9.9 displays a rupture of the film. Such an incidence happened frequently and it meant the failure of the simulation. In this case, the Reynolds' number was 80 and the material properties of water at 20 °C were chosen. The initial film thickness was $\delta_0 = 0.2916\,\text{mm}$ and the time step had a value of $\Delta t = 7.04 \cdot 10^{-8}\,\text{s}$. The plate was $L = 0.175\,\text{m}$ high, divided into 1500 cells. The Weber number was 3.28 and a safety factor of 1.3 for N_{xlimit} was chosen. The excitation frequency had a value of 13 Hz, the amplitude of the perturbation was again 3 %. The rupture took place after 2.56 million time steps.

What are the reasons for the frequent film ruptures? What could be improved in the computer program? First, the explicit scheme for the temporal discretization might have been too inaccurate. In the course of three million time steps, the round-off errors can become too large. Furthermore, the central difference scheme for the spatial discretization may be inadequate. It is known that it can cause big oscillations; see, e. g., Versteeg and Malalasekera [144]. Maybe more care on numerical stability should be taken and an adaptive time step size would yield better results. However, these numerical experiments were disappointing and could not be employed for the simulation of larger control volumes.

9.5 Harmonic waves

In Sections 9.1 to 9.3, the surface was smooth. As a further degree of sophistication, harmonic waves are introduced. First, a simple algorithm will be described. Afterwards, a computationally more expensive program is considered.

9.5.1 An efficient algorithm

The idea is to define a wavy falling film of only the height of a wavelength, $L_0 = \lambda_x$. This control volume statically advances with the phase velocity of the wave:

$$u_{\text{ph}} = \frac{g\delta_0^2}{\nu_L} = \frac{\omega}{k_x} = \frac{2\pi f}{2\pi/\lambda_x}, \tag{9.42}$$

where δ_0 is the thickness of the corresponding smooth film and f denotes the frequency of the wave. The flow rate is modulated from which the film thickness follows

$$\begin{aligned} q_u(x, t = 0) &= q_{u,\text{Nusselt}}(1 + \varepsilon \sin(-k_x x)) \\ \Rightarrow \quad \delta(x, t = 0) &= \left(\frac{3\nu_L q_u(x, t = 0)}{g}\right)^{1/3} \end{aligned} \tag{9.43}$$

The parameter ε is the modulation amplitude, a number between 0 and 1. In what follows, many ideas from Section 9.1 are adopted. The harmonic wave is cut into slices. Each slice is treated by a one-dimensional algorithm taking into account the specific film thickness $\delta(x, t)$. The temperature profile is linear, right from the start,

$$T(x, y, t) = T_{\text{wall}} - (T_{\text{wall}} - T_{\text{surface}}(x, t))\frac{y}{\delta(x, t)} \tag{9.44}$$

Only the diffusion of the salt in the y direction is considered:

$$\frac{\partial S}{\partial t} = D_s \frac{\partial^2 S}{\partial y^2} \tag{9.45}$$

Explicit time marching and central differencing in the cross-stream direction yield:

$$\frac{S_{i,j}^{\text{new}} - S_{i,j}}{\Delta t} = D_s \frac{S_{i,j-1} - 2S_{i,j} + S_{i,j+1}}{(\Delta y)_i^2} \tag{9.46}$$

A zero gradient boundary condition has to be imposed at the wall:

$$\left.\frac{\partial S}{\partial y}\right|_{\text{wall}} = 0 \tag{9.47}$$

Note that $\Delta y = \delta(x, t)/\text{Ny}$, where Ny denotes the number of cells in the y direction. At the surface, the salinity is increased depending on the evaporation:

$$S_{i,\text{Ny}-1}^{\text{new}} = S_{i,\text{Ny}-1}^{\text{old}} + S_{\text{in}}(\Delta_e\delta)_i/(\delta_{\text{Nusselt}}/\text{Ny}), \tag{9.48}$$

where

$$\Delta_e\delta = \frac{\lambda|\partial T/\partial y|_{\text{surface}}}{2\rho \cdot \Delta h} \cdot \Delta t. \tag{9.49}$$

Also the boiling point elevation is taken into account employing the function bpe.

$$\begin{aligned} t = 0: \quad & \mathrm{eto}_i = \mathrm{bpe}(T_{\mathrm{surface},t=0}, S_{\mathrm{in}}) \\ t > 0: \quad & \mathrm{etn}_i = \mathrm{bpe}(T_{i,\mathrm{surface},t}, S_{i,\mathrm{Ny}-1,t}) \\ & T_{i,\mathrm{surface},t} = T_{i,\mathrm{surface},t} + \mathrm{etn}_i - \mathrm{eto}_i \\ & \mathrm{eto}_i = \mathrm{etn}_i \end{aligned} \tag{9.50}$$

With the aid of the surface temperature, the gradient is calculated:

$$\left.\frac{\partial T}{\partial y}\right|_{i,\mathrm{surface},t} = \frac{T_{i,\mathrm{surface},t} - T_{i,\mathrm{Ny}-1,t}}{0.5(\delta(x,t)/\mathrm{Ny})} \tag{9.51}$$

A normalization is performed to conserve the amount of salt inside the falling film:

$$\begin{aligned} t = 0: \quad & \mathrm{fnorm}_i = \delta_i S_{\mathrm{in}} \\ t > 0: \quad & \mathrm{fint}_i = \frac{\delta_i}{\mathrm{Ny}} \sum_{j=0}^{\mathrm{Ny}-1} S_{i,j} \\ & S_{i,j} = S_{\mathrm{in}} + (S_{i,j} - S_{\mathrm{in}})\mathrm{fnorm}_i/\mathrm{fint}_i \end{aligned} \tag{9.52}$$

If L is the height of the heating plate, the harmonic wave reaches its bottom after a total time of

$$t_{\mathrm{total}} = L/u_{\mathrm{ph}}. \tag{9.53}$$

In the calculation of the total simulated time, other choices for the streamwise velocity are possible, of course. The maximal velocity of the corresponding smooth film may rather be chosen, or its average velocity. The result of the simulation is an estimation and has to be tuned anyway; see Section 9.7. Since the explicit time scheme is used, there is an upper limit for the size of the time step:

$$\Delta t < \min_i \left(\frac{(\Delta y)_i^2}{2D_s} \right) \tag{9.54}$$

9.5.2 A more complex program

This program rather imitates a two-dimensional direct numerical simulation. At $x = 0$, there is an inlet where the flow rate is modulated harmonically:

$$\begin{aligned} & q_u(x=0,t) = q_{u,\mathrm{Nusselt}}(1 + \varepsilon \sin(\omega t)) \\ \Rightarrow \quad & \delta(x=0,t) = \left(\frac{3\nu_L q_u(x=0,t)}{g} \right)^{1/3} \end{aligned} \tag{9.55}$$

Let us first have a look on how the film thickness δ is determined. There are Nx values of it. The index i runs from 0 to Nx−1. The wave equation of a harmonic wave, equation (4.72), relates the temporal change of δ with its spatial change:

$$\left(\frac{\partial\delta}{\partial t}\right)_i = -u_{\mathrm{ph}}\left(\frac{\partial\delta}{\partial x}\right)_i \tag{9.56}$$

This spatial change may be determined by

$$\begin{aligned}\left(\frac{\partial\delta}{\partial x}\right)_0 &= \frac{\delta_{i=0}-\delta_{x=0}}{0.5\Delta x}\\ \left(\frac{\partial\delta}{\partial x}\right)_{0<i<\mathrm{Nx}-1} &= \frac{\delta_{i+1}-\delta_{i-1}}{2\Delta x}\\ \left(\frac{\partial\delta}{\partial x}\right)_{\mathrm{Nx}-1} &= \frac{\delta_{\mathrm{Nx}-1}-\delta_{\mathrm{Nx}-2}}{\Delta x}\end{aligned} \tag{9.57}$$

The new film thickness follows from the temporal change and the reduction due to the evaporation:

$$\delta_i^{\mathrm{new}} = \delta_i + \frac{\partial\delta}{\partial t}(\Delta t) - \Delta_e\delta \tag{9.58}$$

The temperature gradient at the surface is the driving "force" for the evaporation:

$$(\Delta_e\delta)_i = \frac{\lambda_L(T_{\mathrm{wall}} - T_{i,\mathrm{surface}})}{2\rho_L\cdot\Delta h\cdot\delta_i}\cdot\Delta t. \tag{9.59}$$

To determine the salinity S, the streamwise convection and the cross-stream diffusion are taken into account:

$$\frac{\partial S}{\partial t} + u\frac{\partial S}{\partial x} = D_s\frac{\partial^2 S}{\partial y^2} \tag{9.60}$$

Using an adaptive grid actually involves using curvilinear coordinates. We would have to write the convection-diffusion equation in tensor form. It would be necessary to know about contravariance, covariance, and the concept of a metric. This would lead too far in a monograph mainly for chemical engineers. A nice document on the Internet about the Laplace operator in curvilinear coordinates is [61].

Since the salinity depends on x and y, we have the two indices i and j. The latter runs from 0 to Ny − 1. The updated salinity in the numerical program can be calculated by

$$S_{i,j}^{\mathrm{new}} = S_{i,j} + (\Delta t)\left[-u_{i,j}\left(\frac{\partial S}{\partial x}\right)_{i,j} + D_s\left(\frac{\partial^2 S}{\partial y^2}\right)_{i,j}\right]. \tag{9.61}$$

The spatial derivatives are calculated according to equation (9.36), mainly. Special care has to be taken at the boundaries:

$$\begin{aligned}\left(\frac{\partial S}{\partial x}\right)_{0,j} &= \frac{S_{i=0,j} - S_{\text{in}}}{0.5\Delta x}\\ \left(\frac{\partial S}{\partial x}\right)_{0<i<\text{Nx}-1,j} &= \frac{S_{i+1,j} - S_{i-1,j}}{2\Delta x}\\ \left(\frac{\partial S}{\partial x}\right)_{\text{Nx}-1,j} &= \frac{S_{\text{Nx}-1,j} - S_{\text{Nx}-2,j}}{\Delta x}\end{aligned} \tag{9.62}$$

$$\begin{aligned}\left(\frac{\partial^2 S}{\partial y^2}\right)_{i,0} &= \frac{S_{i,1} - S_{i,0}}{\Delta y_i^2}\\ \left(\frac{\partial^2 S}{\partial y^2}\right)_{i,0<j<\text{Ny}-1} &= \frac{S_{i,j-1} - 2S_{i,j} + S_{i,j+1}}{\Delta y_i^2}\\ \left(\frac{\partial^2 S}{\partial y^2}\right)_{i,\text{Ny}-1} &= \frac{S_{i,\text{Ny}-2} - S_{i,\text{Ny}-1}}{\Delta y_i^2}\end{aligned} \tag{9.63}$$

$$S_{i,\text{Ny}-1}^{\text{new}} = S_{i,\text{Ny}-1}^{\text{new}} + S_{\text{in}}(\Delta_e\delta)_i/(\delta_{\text{Nusselt}}/\text{Ny}) \tag{9.64}$$

The velocity is assumed to be unidirectional and to have a Nusselt profile.

$$u_{i,j} = \frac{g}{\nu_L}(\delta_i)^2\left[\frac{j}{\text{Ny}-1} - \frac{1}{2}\left(\frac{j}{\text{Ny}-1}\right)^2\right] \tag{9.65}$$

The increasing salinity causes a higher temperature at the surface:

$$\begin{aligned} t = 0: \quad & \text{eto}_i = \text{bpe}(T_{\text{surface},t=0}, S_{\text{in}})\\ t > 0: \quad & \text{etn}_i = \text{bpe}(T_{i,\text{surface},t}, S_{i,\text{Ny}-1,t})\\ & T_{i,\text{surface},t} = T_{i,\text{surface},t} + \text{etn}_i - \text{eto}_i\\ & \text{eto}_i = \text{etn}_i\end{aligned} \tag{9.66}$$

A normalization is performed to conserve the amount of salt inside the falling film:

$$\begin{aligned} t = 0: \quad & \text{fnorm}_i = \delta_i S_{\text{in}}\\ t > 0: \quad & \text{fint}_i = \frac{\delta_i}{\text{Ny}}\sum_{j=0}^{\text{Ny}-1} S_{i,j}\\ & S_{i,j} = S_{\text{in}} + (S_{i,j} - S_{\text{in}})\text{fnorm}_i/\text{fint}_i\end{aligned} \tag{9.67}$$

Figure 9.10 shows the interface of an evaporating harmonic wave. The Reynolds' number is 220, the wall temperature $T_w = 65\,°\text{C}$, the initial saturation temperature at the surface $T_s = 60\,°\text{C}$. The length of the control volume is $L = 0.147\,\text{m} = 600\delta_0$.The excitation amplitude ε for q_u is 10 %, the frequency 40 Hz. The surface is shown after a simulated time of 0.26 s. The effect of the evaporation can be clearly seen, the film thickness is decreasing.

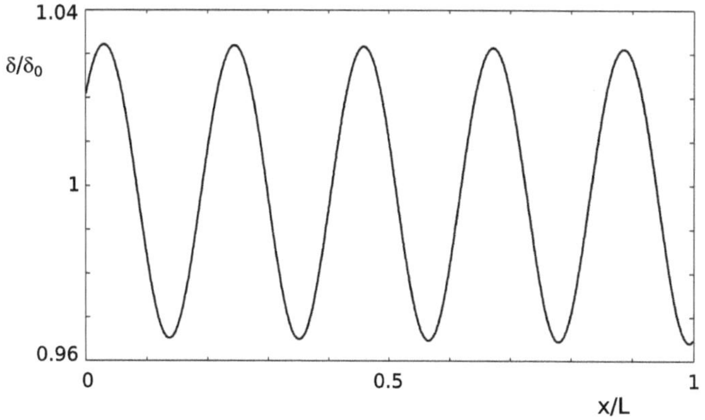

Figure 9.10: Evaporating harmonic wave.

9.6 With input from OpenFOAM

A direct numerical simulation by OpenFOAM can serve as input. Of primary interest is the film thickness δ which can be obtained from c_{VOF}. Within the application, interFoam c_{VOF} is called "alpha.water." With the first component of the velocity, u, the streamwise specific flow rate q_u can be determined. Thickness and flow rate can be used in a two-dimensional simulation of the convection-diffusion type. If only δ is employed, slices of the control volume can be treated one-dimensionally. First, the read-out will be described in Section 9.6.1 which is not trivial. In Section 9.6.2, only the diffusion equation is solved, whereas in Section 9.6.3 the streamwise convection is also considered. Finally, it will be shown how to randomize the excitation at the inlet or how to make it more irregular, at least.

9.6.1 Reading out

From our interFoam case, we copy the files `alpha.water` and `U` of the same directory, e. g., `0.7`, into a new directory. We are interested in the internal fields. We have to delete all the lines that do not contain the wanted numbers. The edited file `alpha.water` looks like this:

```
1
1
1.00001
1
1.00002
1.00002
0.999979
```

```
1
1.00002
0.999993
//..
```

The file U should only contain the velocity vectors:

```
(0.055606 0.0083446 0)
(0.0259109 0.00274142 0)
(0.0120757 -0.000153827 0)
(0.0129824 -0.000284029 0)
(0.0141519 0.00166849 0)
//..
```

In the following, an excerpt of a read-out program is listed. It should be self-explanatory, at least for those who know C++. The character variable bra reads the left bracket of the vector, the variable ket the right bracket. This is necessary, since the brackets are not white space.

```
//first read alpha
for (int j=0;j<Ny;j++){
  for (int i=0;i<Nx;i++){
    alpha >> cvof[i][j];
  }
}
//now read U
char bra,ket;
for (int j=0;j<Ny;j++){
  for (int i=0;i<Nx;i++){
    Vel >> bra;
    Vel >> Ux[i][j];
    Vel >> Uy[i][j];
    Vel >> Uz[i][j];
    Vel >> ket;
  }
}
//determine delta and q
double delta[Nx],q[Nx];
for (int i=0;i<Nx;i++){
  delta[i]=0.0;
  q[i]=0.0;
  for (int j=0;j<Ny;j++){
```

```
      delta[i]+=cvof[i][j]*dy;
      q[i]+=Ux[i][j]*cvof[i][j]*dy;
    }
}
```

Afterwards, only a part of the film may serve as input.

9.6.2 Without convection

The procedure can be analogous to Section 9.5.1 with the difference that the film thickness $\delta(x)$ has been obtained from OpenFOAM. Figure 9.11 shows the result of such a simulation. The input length is $L_0 = 0.035$ m, the simulated time 1.48 s, which corresponds to a length $L = 0.5$ m if the conversion factor is the velocity of an harmonic wave. Re $= 30$, $T_w = 65\,°C$, $T_s = 60\,°C$, $f = 20$ Hz, $\varepsilon = 0.03$. The dark line is the thickness at the inlet, the grey line marks δ at the bottom of the plate. In Figure 9.12, the averaged relative reduction of the film thickness is shown versus the position of the heating plate. The graph is approximately a straight line. Its slope is a measure for the evaporation rate. It can be seen from Figure 9.11 that δ decreases rapidly especially where the film is thin. At such a location, a rupture occurs which prevents the simulation of high plates with this method. A workaround may be to simulate a rather short plate and to extrapolate the straight line in a plot like Figure 9.12. Furthermore, it should be mentioned that the slicing method leads to very high values of the salinity in the troughs. The salt diffusion is only simulated in the cross-stream direction. In the example, a maximal surface salinity of 2.5 in units of the initial salinity S_{in} is

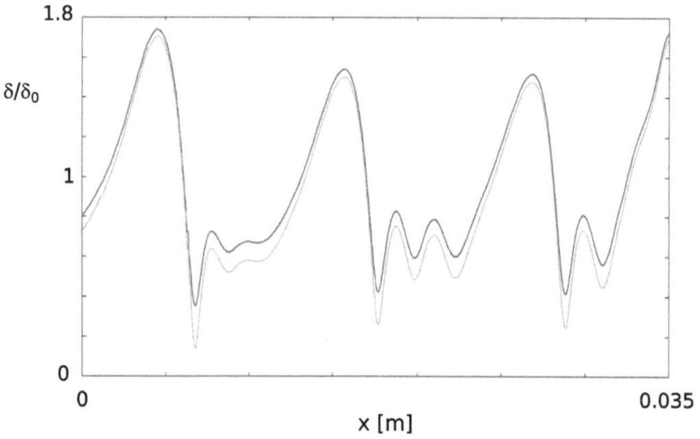

Figure 9.11: Evaporating wavy film (input from OpenFOAM). Here, the streamwise coordinate x serves as a measure for the length of the control volume. The dark line marks the surface of the film at the top of the heating plate. The grey line shows the surface at the bottom of the plate.

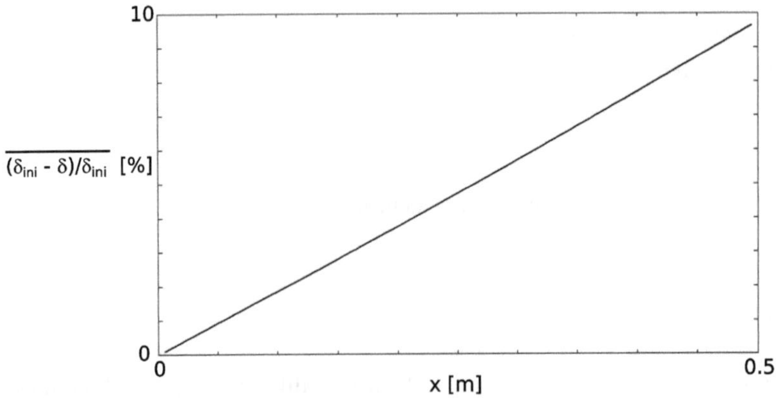

Figure 9.12: The averaged reduction of the film thickness versus the distance from the inlet (input from OpenFOAM).

reached. This value is lower if also the diffusion in the streamwise direction is taken into account.

How much of the film is evaporated in the example of Figure 9.11? After 1.5 s, 9.8 % of the film evaporated on average. If the film were smooth, the relative reduction of the film thickness would be 6.1 %. In the case of pure water, 7.1 % of a film without waves would evaporate.

9.6.3 With convection

Only the outline of an idea is presented here. It can be put into practice with the techniques detailed before. The control volume consists of a part of the output from OpenFOAM. The length is L_0. The film thickness δ and the field of the streamwise velocity $u(x, y)$ are known. The algorithm is very similar to the one of Section 9.6.2 with the exception that the convection is taken into account for the salinity:

$$\frac{\partial S}{\partial t} + u\frac{\partial S}{\partial x} = D_s \frac{\partial^2 S}{\partial y^2} \tag{9.68}$$

The control volume is transported rigidly along the length L of the heating plate. A cyclic boundary condition is imposed for the salinity:

$$S_{0,j} = S_{\mathrm{Nx}-1,j} \quad 0 \le j \le \mathrm{Ny} - 1 \tag{9.69}$$

The specific flow rate $q_u(x)$ could be used instead of $u(x, y)$ from OpenFOAM. In this case, the velocity field would be

$$u(x,y) = \frac{3q_u(x)}{\delta(x)}\left[\frac{y}{\delta(x)} - \frac{1}{2}\left(\frac{y}{\delta(x)}\right)^2\right]. \tag{9.70}$$

9.6.4 Random excitation

In this section, advice is given how to excite the film in another way. Up to now, all excitations were monochromatic harmonic. In the following, an example is given how a random number is generated within a C++ program.

```
#include<cstdlib>
//...
rnum=rand()/(RAND_MAX + 1.);
rn2=rand()/(RAND_MAX + 1.);

if (rn2<=0.5){
  vorz=1.0;
}
else {
  vorz=-1.0;
}

u=umnu*(1.+vorz*rnum*epsilon);
```

The division by `RAND_MAX + 1.` is done to have `rnum` and `rn2` in the open interval between 0 and 1. With `rn2`, the sign is chosen (English "sign," German "Vorzeichen"). The variable `umnu` is the mean velocity according to Nusselt.

Other ways to make the input more irregular are possible, of course. For example, more than one angular frequency can serve as input:

$$\begin{aligned} q_u(x=0,t) &= q_{u,\text{Nusselt}}(1+\varepsilon_1 \sin(\omega_1 t) + \varepsilon_2 \sin(\omega_2 t) + \varepsilon_3 \sin(\omega_3 t)) \\ \Rightarrow \quad \delta(x=0,t) &= \left(\frac{3\nu_L q_u(x=0,t)}{g}\right)^{1/3} \end{aligned} \tag{9.71}$$

The sinusoidal functions may also be multiplied with each other:

$$q_u(x=0,t) = q_{u,\text{Nusselt}}(1+\varepsilon \sin(\omega_1 t)\sin(\omega_2 t)\sin(\omega_3 t)) \tag{9.72}$$

9.7 The evaporation rate

A definition of the evaporation rate can be

$$r_{\text{evap}} = \overline{\frac{\delta_{t=0} - \delta_t}{tLZ}}. \tag{9.73}$$

It is the average reduction of the film thickness per time per area of the heating plate. (The evaporation rate may also be defined differently. In Section 2.2.3 on the EasyMED

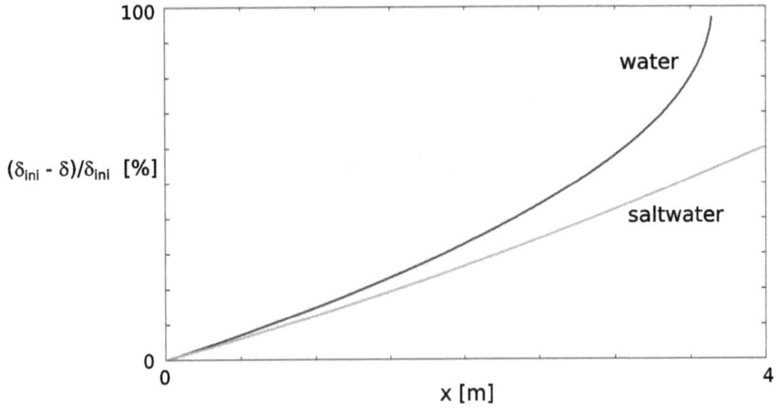

Figure 9.13: Relative reduction of film thickness due to evaporation versus distance from the inlet. The simulation was 1D for a film of Re = 30 and T = 62 °C. For pure water, there is a rupture after about 3.5 m. The solute in the saltwater film moderates the evaporation.

project, it was the relative reduction of the feed.) This quantity is very important in desalination, for it is a measure for the potable water to gain. Figure 9.13 shows the relative reduction of the film thickness due to evaporation versus the distance from the inlet. The simulation was one-dimensional for a film of Re = 30 and $T = 62\,°C$. For pure water, there is a rupture after about 3.5 m. The solute in the saltwater film moderates the evaporation. From a figure like this, the evaporation rate can be deduced. In Figure 9.13, the abscissa actually is the time t according to $x/t = u$. For this figure, u was chosen the velocity of a harmonic wave which is three times the average velocity in a smooth film.

Why is pure water evaporating faster? Recall that the temperature gradient at the surface is the driving force. In the 1D model, it is calculated from

$$\frac{\partial T}{\partial y} = \frac{T_s(S) - T_w}{\delta}, \tag{9.74}$$

where $T_s(S)$ is the surface temperature depending on the salinity. As the salinity increases there, the saturation temperature increases, too. On the other hand, the gradient is rising on account of the decreasing δ. In the case of pure water, the temperature difference in equation (9.74) is constant in the one-dimensional simulation. Therefore, r_{evap} does not rise so much in the case of saltwater.

What is meant by "the results of the original programs have to be tuned anyway?" The prediction of these programs have to be interpreted as approximations. The best way to use them is for parameter studies. For example, the evaporation rate versus the Reynolds' number can be examined. It is also easy to calculate with another wall temperature or a different initial salinity. If experimental data are available, the tuning may look like this:

$$r_{\text{evap},2}^{\text{estimated}} = r_{\text{evap},1}^{\text{measured}} \frac{r_{\text{evap},2}^{\text{CFD}}}{r_{\text{evap},1}^{\text{CFD}}} \tag{9.75}$$

The evaporation rate for a certain condition 1 is known from measurements. It is to be estimated for another condition 2. Both rates are calculated by CFD.

10 Generalization

This book hopefully serves as an inspiration for many engineers not only dealing with falling films but with evaporating solvents in general. In this chapter, we concentrate on original programs. In principle, the methods of Chapter 9 can also be applied to a liquid film that is not necessarily falling. We can keep most of the notation. Figure 10.1 shows a possible set-up of a case. The gravitational acceleration is only requested not to point away from the heating wall. We choose the coordinate system such that u is the unidirectional velocity. This means that x is the streamwise coordinate, y the cross-stream coordinate. The spanwise width is denoted by Z. We essentially treat the problem two-dimensionally, Z will soon be omitted. The thickness of the film still is δ. Throughout this book, the wall temperature is assumed to be constant. Another option would be a constant heat flux which would require a fixed gradient. In Chapter 9, we saw how to implement a zero gradient boundary condition. From there it is straightforward to impose a non-vanishing gradient at the boundary. Almost everything that was described in Chapter 9 can be adopted one to one here, except for the evaporation where a Nusselt velocity profile was assumed. How can the formula for $\Delta_e\delta$ be

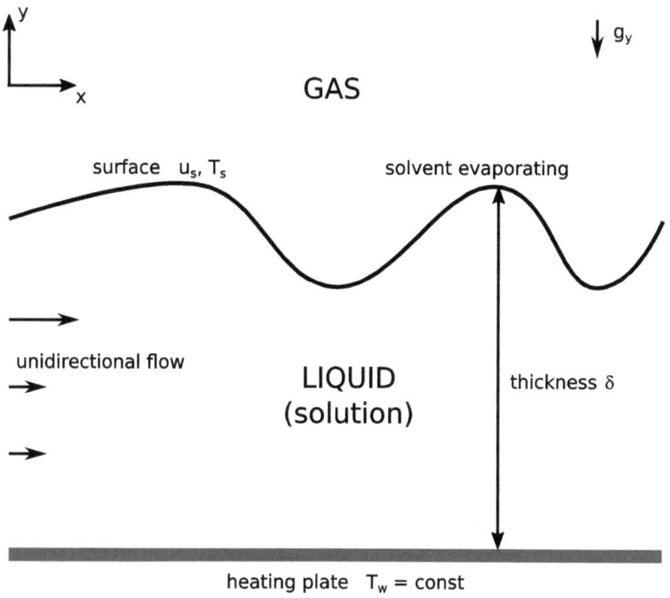

Figure 10.1: Generalization set-up.

https://doi.org/10.1515/9783110592337-010

generalized? The answer is equation (10.1), which will be proved in the following:

$$\boxed{\Delta_e\delta = \frac{\lambda_L|\partial T/\partial y|_s}{\rho_L\,\Delta h f_s\,(1 + (du_s/d\delta)(\delta/u_s))}\cdot\Delta t} \tag{10.1}$$

The subscript "s" indicates the surface of the film.

For a proof, first recall the equations (4.20) and (4.21):

$$d\dot{Q} = \Delta h\, d\dot{M} \tag{10.2}$$

$$d\dot{Q} = \lambda_L\frac{T_w - T_{\text{sat}}}{\delta}Z\,dx = \lambda_L|\partial T/\partial y|_s Z\,dx \tag{10.3}$$

The temporal change of the mass is related to the flow rate:

$$\dot{M} = \rho_L\, Z\, q_u \quad \Rightarrow \quad d\dot{M} = \rho_L\, Z\, dq_u \tag{10.4}$$

Let us assume that the average velocity $\overline{u}$ is proportional to the surface velocity:

$$q_u = \overline{u}\delta = f_s u_s\delta; \quad 0 < f_s < 1. \tag{10.5}$$

In the case of a falling film with a semiparabolic velocity profile, $f_s = 2/3$; cf. equation (4.83). The differential of q_u is

$$dq_u = \frac{dq_u}{d\delta}d\delta = f_s\left(\frac{du_s}{d\delta}\delta + u_s\right)d\delta \tag{10.6}$$

Combining the equations (10.2), (10.3), (10.4), and (10.6) yield

$$\begin{aligned} \Delta h\,\rho_L Z f_s\left(\frac{du_s}{d\delta}\delta + u_s\right)d\delta &= \lambda_L|\partial T/\partial y|_s Z\,dx \\ \Rightarrow \quad d\delta &= \frac{\lambda_L}{\rho_L\,\Delta h}\left|\frac{\partial T}{\partial y}\right|_s\frac{dx}{f_s(u_s + \delta(du_s/d\delta))} \end{aligned} \tag{10.7}$$

We would like to have $d\delta$ depending on the time step dt. Therefore, we make use of the chain rule:

$$dx = \frac{dx}{dt}dt = u_s\,dt \tag{10.8}$$

In the last step, we are taking into account that the evaporation takes place at the surface. Substituting dx into equation (10.7) yields

$$d\delta = \frac{\lambda_L}{\rho_L\,\Delta h}\left|\frac{\partial T}{\partial y}\right|_s\frac{u_s}{f_s(u_s + \delta(du_s/d\delta))}\cdot dt \tag{10.9}$$

In a numerical algorithm, the differentials have a certain size: $d\delta \to \Delta\delta$ and $dt \to \Delta t$. Thus we have proved equation (10.1). As an exercise, the reader may check that equation (9.30) is obtained in case of a Nusselt velocity profile, equation (4.86).

Let us also treat the case where f_s in equation (10.5) is not a constant. In equation (10.6), we could not express dq_u in terms of u_s. Equation (10.1) would become

$$\boxed{\Delta_e\delta = \frac{\lambda_L}{\rho_L\,\Delta h}\left|\frac{\partial T}{\partial y}\right|_s \frac{u_s}{dq_u/d\delta}\cdot\Delta t.} \tag{10.10}$$

What if $dq_u/d\delta$ or $du_s/d\delta$ are not available? Again, the chain rule may be of help:

$$\frac{dq_u}{d\delta} = \frac{dq_u}{dx}\frac{1}{d\delta/dx} \quad \text{and} \quad \frac{du_s}{d\delta} = \frac{du_s}{dx}\frac{1}{d\delta/dx} \tag{10.11}$$

We have almost arrived at the end of this monograph, which will be discussed and its main results highlighted in the last chapter.

11 Discussion and conclusion

Sometimes a calculation engineer hears from an experimentalist, "You regard your simulations as more real than real!" Of course, they are not. Although it is still difficult to simulate evaporating seawater falling films, this book hopefully has conveyed fundamentals and a wealth of insights, new ideas, original programs, and practical advice. In this summary, first open questions shall not be concealed. Then some accomplishments shall be highlighted.

Yet, so many aspects of heated falling liquid films have not been treated. *What about dry-out?* This would involve dry areas on the heating plate and lines where the gas, liquid, and solid phase would meet. In CFD, a constant contact angle is usually used in these cases. However, this is only an approximation. Furthermore, before the dry-out, the liquid film would consist of only a few atomic layers. Thus, the continuum hypothesis, the foundation of fluid dynamics, is violated. The interested reader is referred to Shikhmurzaev [129] to get an impression of the complexity of this matter. *What about turbulent falling liquid films?* The nature of fluid turbulence is still a mystery to physics and engineering. For a recent conceptual account, see e. g., Tsinober's 2nd edition [138]. However, in this treatise a turbulence model has been proposed. Furthermore, many other models were cited and outlined. *What about the salt in the water?* The influence of the salt on the water properties has been modeled in original programs. It was not simulated within OpenFOAM. *What about evaporation in wavy films?* Also this phenomenon has not been simulated within OpenFOAM for this treatise. These matters were at least approximatively considered in the original programs. *Was the control volumes long enough in the simulations with OpenFOAM?* Åkesjö et al. recently published a paper [6] where they report on natural frequencies observed experimentally in very high set-ups. They recommend a control volume not shorter than $1000\delta_0$. In Chapter 7, it was not longer than $600\delta_0$. With the aid of the annexes B, C, and D, the reader is encouraged to perform her/his own simulations making use of the increasing computer power. *What about the gases?* Where would this lead to? If the influence of the gases on the evaporation of the water should be included into the simulation, this would require a one to one modeling. Every detail of the experimental set-up would have to be mapped which is not possible with the present computers. It would be taking a sledgehammer to crack a nut.

The case of evaporating saltwater films around turbulence wires was studied in depth with commercial and open CFD software. In the commercial software, the effect of the surface tension could not be simulated. However, with the open CFD software, it was proved that the main result was correct. With the aid of self-written computer programs, the whole falling film was captured, assuming a film which is not disturbed by turbulence wires. In these programs, our own new models concerning the heat and mass transfer in an evaporating saltwater film are employed. Many details of these programs were described.

https://doi.org/10.1515/9783110592337-011

Let me briefly summarize the main results:

- The distance between turbulence wires should be about 18 wire diameters.
- The OpenFOAM application interFoam is well suited for laminar, wavy falling water films.
- A new application, thinter = thermal interFoam, has given insights into the heat transfer inside a wavy falling film.
- Evaporating falling saltwater films can be simulated by 1D programs under the assumption that they have a smooth surface.
- The conjugate heat transfer from the condensing film through the metal plate and the evaporating seawater film is simulated by a fast 1D program.
- It was shown how to refine the algorithms with a smooth surface to wavy falling films. As a first step, harmonic waves were considered. Then the results of a direct numerical simulation by OpenFOAM served as input.
- The salinity at the surface can be estimated.
- The elevation of the boiling point can be taken into account.
- A formula for the reduction of the film thickness due to evaporation has been found.
- A generalization of the evaporation formula is possible.

A physicist may be called “a physicist’s physicist.” An engineer may respectfully be referred to as “an engineer’s engineer.” With this monograph, I would rather prefer to be “an engineer’s physicist.”

A Proof of equation (9.30)

The following assertion is to be proved in this Appendix:

$$\Delta_e\delta = \frac{\lambda|\partial T/\partial y|_i}{2\rho \cdot \Delta h} \cdot \Delta t \tag{A.1}$$

Recall the equations (4.20), (4.21), and (4.22):

$$d\dot{Q} = \Delta h \, d\dot{M} \tag{A.2}$$

$$d\dot{Q} = \lambda_L \frac{T_W - T_{\text{sat}}}{\delta} Z \, dx \tag{A.3}$$

$$\frac{d\dot{M}}{d\delta} = \frac{\rho_L(\rho_L - \rho_G)}{\mu_L} gZ\delta^2 \tag{A.4}$$

However, equation (A.3) is only valid in the case of a linear temperature profile. Therefore, we should rather employ the temperature gradient at the interface:

$$d\dot{Q} = \lambda_L \left| \frac{\partial T}{\partial y} \right|_i Z \, dx \tag{A.5}$$

Furthermore, in equation (A.4) we neglect ρ_G in comparison to ρ_L:

$$\frac{d\dot{M}}{d\delta} \approx \frac{\rho_L^2}{\mu_L} gZ\delta^2 \tag{A.6}$$

Combining equations (A.2), (A.5), and (A.6) give

$$\left(\rho^2 \cdot g \cdot Z \cdot \delta^2 \cdot d\delta \big/ \mu\right) = \left(\lambda \left| \frac{\partial T}{\partial y} \right|_i \cdot Z \cdot \frac{dx}{dt} \cdot dt \Big/ \Delta h\right). \tag{A.7}$$

In equation (A.7), the index "L" has been omitted. Since the evaporation takes place at the interface

$$\frac{dx}{dt} = u_I = u_{\max} = \frac{1}{2}\frac{\rho}{\mu} \cdot g \cdot \delta^2 \tag{A.8}$$

Inserting equation (A.8) into equation (A.7), we obtain

$$d\delta = \frac{\lambda|\partial T/\partial y|_i}{2\rho \cdot \Delta h} \cdot dt. \tag{A.9}$$

Interpreting the differentials $d\delta$ and dt as differences in a numerical algorithm leads to the assertion.

https://doi.org/10.1515/9783110592337-012

B interFoam

An example shall be given how to simulate an isothermal falling film with the OpenFOAM application interFoam. The Reynolds' number shall be 60, the water temperature 62 °C. The inlet velocity shall be read from an input file. Thus the film is excited with a frequency of 20 Hz. The perturbation amplitude is 3 %.

Let us start and create a directory `fallingfilms` and within this directory another directory `Re60T62Hz20`. There we need the three directories `0`, `constant`, and `system`. (It is more convenient to copy them from the tutorial `damBreak` and then to edit them.)

B.1 The directory constant

In the file `g`, we set the gravitational acceleration:

```
value (9.81 0 0);
```

In the file `turbulenceProperties`, we choose the option

```
simulationType laminar;
```

In the file `transportProperties`, the values for the viscosities, the densities, and the surface tension are set:

```
phases (water air);
water
{   transportModel  Newtonian;
    nu              4.86578e-07;
    rho             1007.38; }
air
{   transportModel  Newtonian;
    nu              1.48e-05;
    rho             1; }
sigma            0.06524;
```

(White spaces are often deleted for the sake of brevity.)

B.2 The directory system

The numerical grid is defined in the file `blockMeshDict`. Note that the `scale` is the film thickness of the corresponding smooth film, δ_0.

https://doi.org/10.1515/9783110592337-013

```
scale   0.000163169;
vertices
(   (0 0 0)
    (600 0 0)
    (600 4 0)
    (0 4 0)
    (0 0 1)
    (600 0 1)
    (600 4 1)
    (0 4 1) );
blocks
(  hex (0 1 2 3 4 5 6 7) (1500 40 1) simpleGrading (1 1 1) );
edges
( );
boundary
(   inlet
    {   type inlet;
        faces
        (   (0 4 7 3)  );  }
    outlet
    {   type outlet;
        faces
        (   (2 6 5 1)  );  }
    Wbelow
    {   type wall;
        faces
        (   (1 5 4 0)  );  }
    atmosphere
    {   type patch;
        faces
        (   (3 7 6 2)  );  } );
mergePatchPairs
( );
```

The two files fvSolution and fvSchemes have been taken from the damBreak tutorial without changes. On the one hand, they may change in the course of the years as interFoam will presumably be developed further. On the other hand, it is good to document them. This is the reason why these files are listed here. The file fvSolution reads

```
solvers
{  "alpha.water.*"
   {    nAlphaCorr      2;
```

```
        nAlphaSubCycles 1;
        cAlpha          1;

        MULESCorr       yes;
        nLimiterIter    5;

        solver          smoothSolver;
        smoother        symGaussSeidel;
        tolerance       1e-8;
        relTol          0; }

    "pcorr.*"
    {   solver          PCG;
        preconditioner  DIC;
        tolerance       1e-5;
        relTol          0; }

    p_rgh
    {   solver          PCG;
        preconditioner  DIC;
        tolerance       1e-07;
        relTol          0.05; }

    p_rghFinal
    {   $p_rgh;
        relTol          0; }

    U
    {   solver          smoothSolver;
        smoother        symGaussSeidel;
        tolerance       1e-06;
        relTol          0; }
}

PIMPLE
{   momentumPredictor   no;
    nOuterCorrectors    1;
    nCorrectors         3;
    nNonOrthogonalCorrectors 0; }

relaxationFactors
{   equations
```

```
    {   ".*" 1; }
}
```

In the file fvSchemes, we have

```
ddtSchemes
{   default          Euler; }

gradSchemes
{   default          Gauss linear; }

divSchemes
{   div(rhoPhi,U)  Gauss linearUpwind grad(U);
    div(phi,alpha)  Gauss vanLeer;
    div(phirb,alpha) Gauss linear;
    div(((rho*nuEff)*dev2(T(grad(U))))) Gauss linear; }

laplacianSchemes
{   default          Gauss linear corrected; }

interpolationSchemes
{   default          linear; }

snGradSchemes
{   default          corrected; }
```

The file setFieldsDict is needed to set the initial values for c_{VOF} which is called alpha.water:

```
defaultFieldValues
(   volScalarFieldValue alpha.water 0 );
regions
(   boxToCell
    {   box (0 0 -1) (0.0979011 0.000163169 1);
        fieldValues
        (   volScalarFieldValue alpha.water 1  );  } );
```

The file controlDict reads

```
application     interFoam;
startFrom       startTime;
startTime       0.0;
```

```
stopAt          endTime;
endTime         0.7;
deltaT          2.4e-5;
writeControl    adjustableRunTime;
writeInterval   0.35;
purgeWrite      0;
writeFormat     ascii;
writePrecision  6;
writeCompression off;
timeFormat      general;
timePrecision   6;
runTimeModifiable yes;
adjustTimeStep  yes;
maxCo           1;
maxAlphaCo      1;
maxDeltaT       1;
```

B.3 The directory 0

In this directory, there are the files `alpha.water`, `alpha.water.orig`, `p_rgh`, `U`, and `tfthRe60T62Hz20.txt`. The last file is the input for the inlet velocity.

Let us first have a closer look into `alpha.water.orig`. There the boundary conditions of c_{VOF} are given. It is copied to `alpha.water` where later the initial values are set by the application `setFields` which reads `../system/setFieldsDict`. First, the control volume is filled with gas: `internalField uniform 0;` This will later be modified by the application `setFields`. Then the boundary condition for the inlet is chosen. In the y direction, the control volume is 40 cells high. Into 10 cells water shall flow, into the remaining 30 cells air.

```
boundaryField
{
    inlet
    {   type                fixedValue;
              value               nonuniform List<scalar>
          40
              (
        1
        1
        1
       ...
        0
```

```
        0
        0
        );
    }
```

For this case, `1` appears 10 times and `0` 30 times. The boundary `outlet` is of the type `zeroGradient`. `Wbelow` has the type `fixedValue` with value `uniform 1`. This condition is very important, for it prevents the falling film from being ripped open, which would cause a failure of the simulation. For the boundary `atmosphere`, we set

```
type            inletOutlet;
inletValue      uniform 0;
value           uniform 0;
```

Finally, the `defaultFaces` have the type `empty`.

Let us turn to the file `p_rgh` for the pressure. Initially, the internal field is zero:

```
internalField uniform 0;
```

The boundary `inlet` is of the type `fixedValue` with the value `uniform 0`. The `outlet` is of the type `zeroGradient`. The `atmosphere` has the same type and value as `inlet`. The `defaultFaces` have the type `empty`.

What remains is the file `U` for the velocities. The initial value everywhere is the mean velocity of the Nusselt film:

```
internalField uniform (0.178924 0 0);
```

Special attention is needed for the `inlet`, since we want interFoam to read an input file.

```
type uniformFixedValue;
uniformValue tableFile;
file "$FOAM_CASE/0/tfthRe60T62Hz20.txt";
outOfBounds clamp;
interpolationScheme linear;

value uniform (0.178924 0 0);
```

We will examine the input file a little bit later. Again, the `outlet` is of the type `zeroGradient`. For the `atmosphere`, we choose

```
type            pressureInletOutletVelocity;
value           uniform (0 0 0);
```

Once again, `defaultFaces` are of the type `empty`.

The input file `tfthRe60T62Hz20.txt` contains a list, a time, and a velocity vector in each row.

```
(
(0 (0.178924 0 0))
(4.8e-05 (0.178956 0 0))
(9.6e-05 (0.178988 0 0))
...
(0.699888 (0.178848 0 0))
(0.699936 (0.17888 0 0))
(0.699984 (0.178913 0 0))
);
```

B.4 Running the program

Go into the directory `fallingfilms`. Then generate the mesh:

```
blockMesh -case Re60T62Hz20
```

Set the initial condition for `alpha.water`:

```
setFields -case Re60T62Hz20
```

Finally, we can start the application:

```
interFoam -case Re60T62Hz20
```

This case takes about 50 minutes (just to give an order of magnitude). The final distribution of alpha.water looks like this if the y coordinate is magnified by a factor of 50. See Figure B.1.

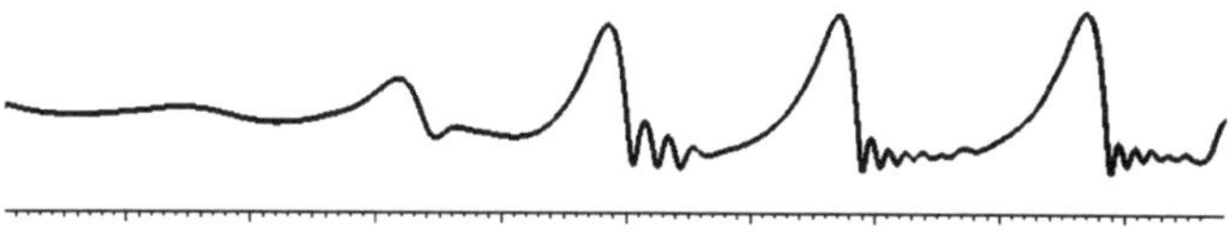

Figure B.1: The interface for Re = 60, $T = 62^{\circ}$C, and a frequency of 20 Hz after 0.7 s.

C thinter

The name of this new application is a shortform of “thermal interFoam.” The existing interFoam is extended by an energy equation. This means that thinter solves also for the temperature. The fluid properties, like density, viscosity, thermal diffusivity, and surface tension, remain constant. They are assumed not to vary too much.

Let us start by creating a directory `thinter`. Then we need to copy some source code into it. The paths that are given may not be identical to the locations on your computer. However, they certainly can be adapted easily. From the directory,

```
/opt/OpenFOAM/OpenFOAM-v1712/applications/solvers/multiphase/interFoam
```

copy the files `alphaSuSp.H`, `createFields.H`, `correctPhi.H`, `interFoam.C`, `pEqn.H`, `rhofs.H`, and `UEqn.H`, and the directory `Make`.

Additionally, copy the directories

```
(/opt/OpenFOAM/OpenFOAM-v1712/src/transportModels/)
immiscibleIncompressibleTwoPhaseMixture
```

and

```
(/opt/OpenFOAM/OpenFOAM-v1712/src/transportModels/incompressible/)
incompressibleTwoPhaseMixture
```

into `thinter`.

The paths in brackets indicate where these directories may be. Change the name of the file `interFoam.C` into `thinter.C`.

Now we have to edit this file:

```
...
   // --- Pressure-velocity PIMPLE corrector loop
   while (pimple.loop())
   {
...
    //NEW:solve for temperature
    solve( fvm::ddt(T) + fvm::div(phi,T) - fvm::laplacian(lamborc,T) );
   }
```

Thus the new application will also solve for the temperature T. The quantity `lamborc` is the short form of “λ over ρ and c_V.” It is the thermal diffusivity.

https://doi.org/10.1515/9783110592337-014

Temperature and diffusivity have a value in each cell of the mesh. These fields have to be read. Therefore, we have to edit the file `createFields.H`. After having read the field `p_rgh`, we read the temperature field:

```
//NEW: read temperature
Info<< "Reading field T\n" << endl;
volScalarField T
(
    IOobject
    (
        "T",
        runTime.timeName(),
        mesh,
        IOobject::MUST_READ,
        IOobject::AUTO_WRITE
    ),
    mesh
);
```

Since there is a two-phase mixture, we have two densities. Behind the densities, we introduce the diffusivities:

```
const dimensionedScalar\& lamborc1 = mixture.lamborc1();
const dimensionedScalar\& lamborc2 = mixture.lamborc2();
```

After `rho` is stored, we do the same with `lamborc`:

```
//NEW: Need to store lamborc for laplacian(lamborc, T)
volScalarField lamborc
(
 IOobject
 (
  "lamborc",
  runTime.timeName(),
  mesh,
  IOobject::READ_IF_PRESENT
  ),
 alpha1*lamborc1 + alpha2*lamborc2
 );
lamborc.oldTime();
```

Now go into the subdirectory `incompressibleTwoPhaseMixture`.

In the header file `incompressibleTwoPhaseMixture.H`, we write new protected data behind `rho1_` and `rho2_`:

```
dimensionedScalar lamborc1_;
dimensionedScalar lamborc2_;
```

Behind the member functions of the densities, we write those of the thermal diffusivities:

```
        //-NEW Return const-access to phase1 thermal diffusivity
        const dimensionedScalar& lamborc1() const
        {
            return lamborc1_;
        }
        //-NEW Return const-access to phase2 thermal diffusivity
        const dimensionedScalar& lamborc2() const
        {
            return lamborc2_;
        };
```

After the header, we edit the file `incompressibleTwoPhaseMixture.C`.

```
rho1_("rho", dimDensity, nuModel1_->viscosityProperties()),
rho2_("rho", dimDensity, nuModel2_->viscosityProperties()),

lamborc1_("lamborc", dimVelocity*dimLength, nuModel1_->viscosityProperties()),
lamborc2_("lamborc", dimVelocity*dimLength, nuModel2_->viscosityProperties()),
```

The first two lines are given here in order to see where `lamborc` is placed. Almost at the end of the file, `rho` is looked up. The same shall be done for `lamborc`:

```
nuModel1_->viscosityProperties().lookup("rho") >> rho1_;
nuModel2_->viscosityProperties().lookup("rho") >> rho2_;
nuModel1_->viscosityProperties().lookup("lamborc") >> lamborc1_;
nuModel2_->viscosityProperties().lookup("lamborc") >> lamborc2_;
```

Again the first two lines are mentioned for redundancy.

C.1 How to compile thinter

Go into the directory `thinter/Make`. The file `files` shall read

```
incompressibleTwoPhaseMixture/incompressibleTwoPhaseMixture.C
immiscibleIncompressibleTwoPhaseMixture/immiscibleIncompressibleTwoPhaseMixture.C
```

```
thinter.C

EXE = $(FOAM_APPBIN)/thinter
```

The font size is smaller to accommodate to the page width. The file `options` shall look like:

```
EXE_INC = \
-I/opt/OpenFOAM/OpenFOAM-v1712/applications/solvers/multiphase/VoF \
-I$(LIB_SRC)/transportModels \
-I$(LIB_SRC)/transportModels/twoPhaseMixture/lnInclude \
-I/mnt/d/Users/Henning/OpenFOAM/thinter/incompressibleTwoPhaseMixture \
-I$(LIB_SRC)/transportModels/interfaceProperties/lnInclude \
-I$(LIB_SRC)/TurbulenceModels/turbulenceModels/lnInclude \
-I$(LIB_SRC)/TurbulenceModels/incompressible/lnInclude \
-I$(LIB_SRC)/transportModels/immiscibleIncompressibleTwoPhaseMixture/lnInclude \
-I$(LIB_SRC)/finiteVolume/lnInclude \
-I$(LIB_SRC)/meshTools/lnInclude \
-I$(LIB_SRC)/sampling/lnInclude

EXE_LIBS = \
-limmiscibleIncompressibleTwoPhaseMixture \
-lturbulenceModels \
-lincompressibleTurbulenceModels \
-lfiniteVolume \
-lfvOptions \
-lmeshTools \
-lsampling \
-lwaveModels
```

We just have to change the file `options` in the directory:

```
thinter/immiscibleIncompressibleTwoPhaseMixture/Make
```

Replace `-I../incompressible/lnInclude`
with:

```
-I../incompressibleTwoPhaseMixture \
```

Finally, we can compile the application thinter. Go into your directory `thinter` and type:

```
wmake
```

In order to check if your compilation has been a successful, type:

```
which thinter
```

The system should give you the full path of the executable, e. g.,

```
/opt/OpenFOAM/OpenFOAM-v1712/platforms/linux64Gcc63DPInt32Opt/bin/thinter
```

C.2 Test case

In Appendix B “interFoam” it was shown how to simulate an isothermal falling film within OpenFOAM. This case shall be extended in order to take the temperature T into account. For this purpose, we go into the directory `fallingfilms` and create a new test case directory `thRe60T62Hz20`. In this directory, we copy `Re60T62Hz20/0`, `Re60T62Hz20/system`, and `Re60T62Hz20/constant`.

In the directory `thRe60T62Hz20/0`, we have to choose initial and boundary conditions in a file with the name `T`:

```
FoamFile
{   version       2.0;
    format        ascii;
    class         volScalarField;
    object        T; }
dimensions        [0 0 0 1 0 0 0];
internalField     uniform 333.15;
boundaryField
{
    inlet
    {   type              fixedValue;
        value             uniform 333.15; }

    outlet
    {   type              zeroGradient;}

    Wbelow
    {   type              fixedValue;
        value             uniform 338.15; }

    atmosphere
    {   type              zeroGradient; }

    defaultFaces
    {   type              empty; }
}
```

(Again, there may be more white spaces which have been deleted for brevity.)

Now change to the directory thRe60T62Hz20/constant and edit the file transportProperties. In addition to nu and rho, the thermal diffusivity lamborc has to be given for the two phases water and air. We choose 1.61e-07 for water and 26.0e-06 in the case of air.

In the directory thRe60T62Hz20/system, we open the file fvSolution. Between the instructions for p_rghFinal and U, we communicate to the application how to solve for T:

```
T
{   solver          smoothSolver;
    smoother        symGaussSeidel;
    tolerance       1e-06;
    relTol          0; }

TFinal
{   $T;
    relTol          0; }
```

In the file fvSchemes under divSchemes, we add:

```
div(phi,T) Gauss linear;
```

Finally, in the file controlDict the startTime is 0, the endTime 0.7, and the writeInterval 0.01. (For a writeInterval of 0.1, the temperature diverged between 0.3 and 0.4 seconds.)

In the directory fallingfilms, we run the test case:

```
thinter -case thRe60T62Hz20
```

It took about 56 minutes. Just to compare a few values of the resulting temperature field, we change into the directory

thRe60T62Hz20/0.7 and open the file T. The internal field should be something like:

```
internalField   nonuniform List<scalar>
60000
(
334.516
336.313
337.062
337.289
337.354
```

```
...
337.17
337.169
337.168
337.167
337.167
)
;
```

Your numbers may be slightly different, since the film thickness and the fluid properties were not exactly the same like in the test case `Re60T62Hz20` of Appendix B “interFoam.”

D Wires

In this Appendix, details will be given how to set up a geometry with turbulence wires within OpenFOAM. Computer aided design (CAD) is not necessary. Figure D.1 shows the blocks of a numerical grid with three wires. The numbers refer to the vertices. Also for a two-dimensional case, there is a third coordinate. That is the reason why there are always two numbers in the x-y-plane, for we have an extension into the z direction. Figure D.1 is only a scheme, the size of the blocks and the distances between the vertices is not drawn to scale.

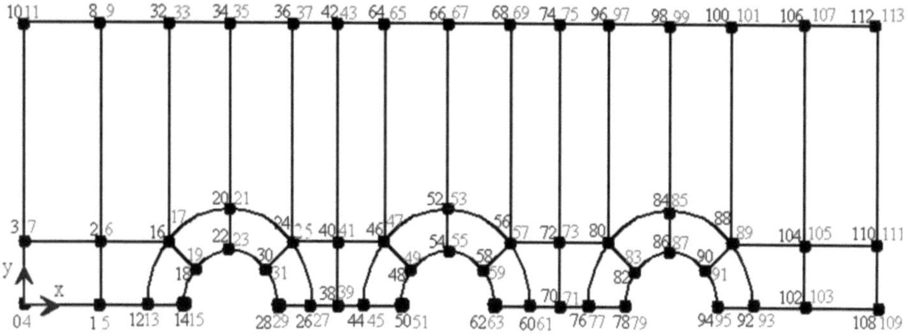

Figure D.1: Blocks of the grid around three wires.

The main difference to the case in Annex C is that the file

```
$FOAM\_CASE/system/blockMeshDict
```

has to be edited. Essentially, the file may read:

```
// excerpt file blockMeshDict

scale   0.000248752;

vertices
(
(0 0 0)
(40 0 0)
(40 1.2727922 0)
(0 1.2727922 0)
(0 0 0.1)
(40 0 0.1)
(40 1.2727922 0.1)
```

https://doi.org/10.1515/9783110592337-015

```
(0 1.2727922 0.1)
(40 4 0)
(40 4 0.1)
(0 4 0)
(0 4 0.1)
(48.2 0 0)
(48.2 0 0.1)
(49.2 0 0)
(49.2 0 0.1)
(48.727208 1.2727922 0)
(48.727208 1.2727922 0.1)
(49.434315 0.5656854 0)
(49.434315 0.5656854 0.1)
(50 1.8 0)
(50 1.8 0.1)
(50 0.8 0)
(50 0.8 0.1)
(51.272792 1.2727922 0)
(51.272792 1.2727922 0.1)
(51.8 0 0)
(51.8 0 0.1)
(50.8 0 0)
(50.8 0 0.1)
(50.565685 0.5656854 0)
(50.565685 0.5656854 0.1)
(48.727208 4 0)
(48.727208 4 0.1)
(50 4 0)
(50 4 0.1)
(51.272792 4 0)
(51.272792 4 0.1)
(56 0 0)
(56 0 0.1)
(56 1.2727922 0)
(56 1.2727922 0.1)
(56 4 0)
(56 4 0.1)
(60.2 0 0)
(60.2 0 0.1)
(60.727208 1.2727922 0)
(60.727208 1.2727922 0.1)
(61.434315 0.5656854 0)
```

```
(61.434315 0.5656854 0.1)
(61.2 0 0)
(61.2 0 0.1)
(62 1.8 0)
(62 1.8 0.1)
(62 0.8 0)
(62 0.8 0.1)
(63.272792 1.2727922 0)
(63.272792 1.2727922 0.1)
(62.565685 0.5656854 0)
(62.565685 0.5656854 0.1)
(63.8 0 0)
(63.8 0 0.1)
(62.8 0 0)
(62.8 0 0.1)
(60.727208 4 0)
(60.727208 4 0.1)
(62 4 0)
(62 4 0.1)
(63.272792 4 0)
(63.272792 4 0.1)
(68 0 0)
(68 0 0.1)
(68 1.2727922 0)
(68 1.2727922 0.1)
(68 4 0)
(68 4 0.1)
(72.2 0 0)
(72.2 0 0.1)
(73.2 0 0)
(73.2 0 0.1)
(72.72720779 1.27279221 0)
(72.72720779 1.27279221 0.1)
(73.43431458 0.565685425 0)
(73.43431458 0.565685425 0.1)
(74 1.8 0)
(74 1.8 0.1)
(74 0.8 0)
(74 0.8 0.1)
(75.27279221 1.27279221 0)
(75.27279221 1.27279221 0.1)
(74.56568543 0.565685425 0)
```

```
(74.56568543 0.565685425 0.1)
(75.8 0 0)
(75.8 0 0.1)
(74.8 0 0)
(74.8 0 0.1)
(72.72720779 4 0)
(72.72720779 4 0.1)
(74 4 0)
(74 4 0.1)
(75.27279221 4 0)
(75.27279221 4 0.1)
(80 0 0)
(80 0 0.1)
(80 1.27279221 0)
(80 1.27279221 0.1)
(80 4 0)
(80 4 0.1)
(600 0 0)
(600 0 0.1)
(600 1.2727922 0)
(600 1.2727922 0.1)
(600 4 0)
(600 4 0.1)
);
blocks
(
hex (0 1 2 3 4 5 6 7) (100 12 1) simpleGrading (1 1 1)
hex (3 2 8 10 7 6 9 11) (100 25 1) simpleGrading (1 1 1)
hex (1 12 16 2 5 13 17 6) (25 12 1) simpleGrading (1 1 1)
hex (2 16 32 8 6 17 33 9) (25 25 1) simpleGrading (1 1 1)
hex (12 14 18 16 13 15 19 17) (10 12 1) simpleGrading (1 1 1)
hex (16 18 22 20 17 19 23 21) (10 5 1) simpleGrading (1 1 1)
hex (20 22 30 24 21 23 31 25) (10 5 1) simpleGrading (1 1 1)
hex (24 30 28 26 25 31 29 27) (10 12 1) simpleGrading (1 1 1)
hex (16 20 34 32 17 21 35 33) (5 25 1) simpleGrading (1 1 1)
hex (20 24 36 34 21 25 37 35) (5 25 1) simpleGrading (1 1 1)
hex (26 38 40 24 27 39 41 25) (25 12 1) simpleGrading (1 1 1)
hex (24 40 42 36 25 41 43 37) (25 25 1) simpleGrading (1 1 1)
hex (38 44 46 40 39 45 47 41) (25 12 1) simpleGrading (1 1 1)
hex (40 46 64 42 41 47 65 43) (25 25 1) simpleGrading (1 1 1)
hex (44 50 48 46 45 51 49 47) (10 12 1) simpleGrading (1 1 1)
hex (46 48 54 52 47 49 55 53) (10 5 1) simpleGrading (1 1 1)
```

```
hex (52 54 58 56 53 55 59 57) (10 5 1) simpleGrading (1 1 1)
hex (56 58 62 60 57 59 63 61) (10 12 1) simpleGrading (1 1 1)
hex (46 52 66 64 47 53 67 65) (5 25 1) simpleGrading (1 1 1)
hex (52 56 68 66 53 57 69 67) (5 25 1) simpleGrading (1 1 1)
hex (60 70 72 56 61 71 73 57) (25 12 1) simpleGrading (1 1 1)
hex (56 72 74 68 57 73 75 69) (25 25 1) simpleGrading (1 1 1)
hex (70 76 80 72 71 77 81 73) (25 12 1) simpleGrading (1 1 1)
hex (72 80 96 74 73 81 97 75) (25 25 1) simpleGrading (1 1 1)
hex (76 78 82 80 77 79 83 81) (10 12 1) simpleGrading (1 1 1)
hex (80 82 86 84 81 83 87 85) (10 5 1) simpleGrading (1 1 1)
hex (84 86 90 88 85 87 91 89) (10 5 1) simpleGrading (1 1 1)
hex (88 90 94 92 89 91 95 93) (10 12 1) simpleGrading (1 1 1)
hex (80 84 98 96 81 85 99 97) (5 25 1) simpleGrading (1 1 1)
hex (84 88 100 98 85 89 101 99) (5 25 1) simpleGrading (1 1 1)
hex (92 102 104 88 93 103 105 89) (25 12 1) simpleGrading (1 1 1)
hex (88 104 106 100 89 105 107 101) (25 25 1) simpleGrading (1 1 1)
hex (102 108 110 104 103 109 111 105) (1300 12 1) simpleGrading (1 1 1)
hex (104 110 112 106 105 111 113 107) (1300 25 1) simpleGrading (1 1 1)
);
edges
(
arc 14 18 (49.26089637 0.306146745 0)
arc 12 16 (48.33701684 0.688830178 0)
arc 18 22 (49.69385325 0.739103626 0)
arc 16 20 (49.31116982 1.662983159 0)
arc 28 30 (50.73910363 0.306146745 0)
arc 26 24 (51.66298316 0.688830178 0)
arc 30 22 (50.30614675 0.739103626 0)
arc 24 20 (50.68883018 1.662983159 0)
arc 15 19 (49.26089637 0.306146745 0.1)
arc 13 17 (48.33701684 0.688830178 0.1)
arc 19 23 (49.69385325 0.739103626 0.1)
arc 17 21 (49.31116982 1.662983159 0.1)
arc 29 31 (50.73910363 0.306146745 0.1)
arc 27 25 (51.66298316 0.688830178 0.1)
arc 31 23 (50.30614675 0.739103626 0.1)
arc 25 21 (50.68883018 1.662983159 0.1)
arc 50 48 (61.26089637 0.306146745 0)
arc 44 46 (60.33701684 0.688830178 0)
arc 48 54 (61.69385325 0.739103626 0)
arc 46 52 (61.31116982 1.662983159 0)
arc 62 58 (62.73910363 0.306146745 0)
```

```
arc 60 56 (63.66298316 0.688830178 0)
arc 58 54 (62.30614675 0.739103626 0)
arc 56 52 (62.68883018 1.662983159 0)
arc 51 49 (61.26089637 0.306146745 0.1)
arc 45 47 (60.33701684 0.688830178 0.1)
arc 49 55 (61.69385325 0.739103626 0.1)
arc 47 53 (61.31116982 1.662983159 0.1)
arc 63 59 (62.73910363 0.306146745 0.1)
arc 61 57 (63.66298316 0.688830178 0.1)
arc 59 55 (62.30614675 0.739103626 0.1)
arc 57 53 (62.68883018 1.662983159 0.1)
arc 78 82 (73.26089637 0.306146745 0)
arc 76 80 (72.33701684 0.688830178 0)
arc 82 86 (73.69385325 0.739103626 0)
arc 80 84 (73.31116982 1.662983159 0)
arc 94 90 (74.73910363 0.306146745 0)
arc 92 88 (75.66298316 0.688830178 0)
arc 90 86 (74.30614675 0.739103626 0)
arc 88 84 (74.68883018 1.662983159 0)
arc 79 83 (73.26089637 0.306146745 0.1)
arc 77 81 (72.33701684 0.688830178 0.1)
arc 83 87 (73.69385325 0.739103626 0.1)
arc 81 85 (73.31116982 1.662983159 0.1)
arc 95 91 (74.73910363 0.306146745 0.1)
arc 93 89 (75.66298316 0.688830178 0.1)
arc 91 87 (74.30614675 0.739103626 0.1)
arc 89 85 (74.68883018 1.662983159 0.1)
);

boundary
(
     inlet
     { type inlet;
       faces
       (
       (0 4 7 3)
       (3 7 11 10)
        );
     }

     outlet
     { type outlet;
```

```
  faces
  (
  (110 111 109 108)
  (112 113 111 110)
   );
}

atmosphere
{ type patch;
  faces
  (
  (10 11 9 8)
  (8 9 33 32)
  (32 33 35 34)
  (34 35 37 36)
  (36 37 43 42)
  (42 43 65 64)
  (64 65 67 66)
  (66 67 69 68)
  (68 69 75 74)
  (74 75 97 96)
  (96 97 99 98)
  (98 99 101 100)
  (100 101 107 106)
  (106 107 113 112)
   );
}

Wbelow
{ type wall;
  faces
  (
  (1 5 4 0)
  (12 13 5 1)
  (14 15 13 12)
  (18 19 15 14)
  (22 23 19 18)
  (30 31 23 22)
  (28 29 31 30)
  (26 27 29 28)
  (38 39 27 26)
  (44 45 39 38)
```

```
        (50 51 45 44)
        (48 49 51 50)
        (54 55 49 48)
        (58 59 55 54)
        (62 63 59 58)
        (60 61 63 62)
        (70 71 61 60)
        (76 77 71 70)
        (78 79 77 76)
        (82 83 79 78)
        (86 87 83 82)
        (90 91 87 86)
        (94 95 91 90)
        (92 93 95 94)
        (102 103 93 92)
        (108 109 103 102)
         );
      }

);


mergePatchPairs
(
);

// end excerpt file blockMeshDict
```

In this case, the `scale` is the thickness of a smooth film with Re = 220 and $T = 60\,°C$.

E Counterstatement

In [37], Deshpande, Anumolu, and Trujillo write:

> Raach et al. [119] implemented an energy equation in the framework of interFoam. The solver was used without surface tension modeling to simulate heat transfer in a film falling over turbulence wires. Validation consisted of comparing phase velocities and wave peak heights with their experiments [118].

(The citation numbers have been adapted to this book.) A similar summary is given by Emad in his Master's thesis [46].

Without wanting to tell which side is to be blamed, I want to correct these misunderstandings:

All simulations done by OpenFOAM were with surface tension. Solely in the earlier numerical experiments with Star-CD, reported in [116], the surface tension had to be set to zero for laminar cases. The experimental values of the phase velocities and wave peak heights were calculated with correlations of Nosoko et al. [105] who had performed real experiments.

https://doi.org/10.1515/9783110592337-016

F List of symbols

This Appendix consists of three tables: Table F.1 explains Greek symbols. Table F.2 lists symbols with Roman letters. Finally, in Table F.3 "mixed" symbols with Greek and Roman letters are made clear.

Table F.1: Table of Greek symbols.

symbol	unit	explanation	see
α	W/m^2/K	heat transfer coefficient	Equation (4.26)
β	–	empirical constant	Equation (3.99)
δ	m	film thickness	
δ_0	m	film thickness of corresponding smooth film	Equation (4.19)
η	m	length scale of smallest turbulent eddies	Equation (3.109)
ε	m^2/s^3	dissipation of turbulent kinetic energy	Equation (3.77)
ε	–	modulation amplitude	Equation (4.73)
Γ	m^3/s	flow rate	Equation (6.1)
Γ_ϕ	depends	material property in model equation	Equation (5.1)
λ	W/m/K	heat conductivity	
μ	Pa s	dynamic viscosity	
ν	m^2/s	kinematic viscosity	
ϕ	depends	physical quantity in model equation	Equation (5.1)
ρ	kg/m^3	density	
σ	N/m	surface tension	
τ	Pa	stress	Figure 3.1
κ_1, κ_2	1/m	principal curvatures	Equation (3.46)
κ	1/m	curvature (= $2H$)	Equation (5.50)
κ	–	Karman's constant	Equation (3.88)
ω	Hz	angular frequency	
Σ_ε	m^2/s^4	other terms of k-ε model	Equation (3.111)
θ	°	inclination angle of plate	Figure 4.6
ξ	m	length scale of largest turbulent eddies	Equation (3.108)

https://doi.org/10.1515/9783110592337-017

Table F.2: Table of symbols with Roman letters.

symbol	unit	explanation	see
A	m^2	area	
A	–	additive constant	Equation (3.88)
a	–	constant	Equation (3.106)
$\boldsymbol{a}$	m/s^2	acceleration	
$\boldsymbol{b}$	$1/m$	surface curvature dyadic	Equation (3.45)
c_p	J/kg/K	specific heat capacity at constant pressure	
c_V	J/kg/K	specific heat capacity at constant volume	
Co	–	Courant number	Equation (5.28)
D/Dt	$1/s$	substantial derivative	Equation (3.5)
D_{th}	m^2/s	thermal diffusivity	Equation (3.32)
$D_{\mathrm{th},t}$	m^2/s	turbulent thermal diffusivity	Equation (3.114)
D_s	m^2/s	salt diffusivity	
D_v	m^2/s	vapor diffusivity	
e	–	Euler's number	
e	m^2/s^2	specific inner energy	
e	–	eastern cell face	Figure 5.1
E	–	eastern neighbor of P	Figure 5.1
$\boldsymbol{F}$	N	force	
$\boldsymbol{F}_{\mathrm{CSF}}$	N/m^3	Laplace volume force in CSF model	Equation (5.53)
$\boldsymbol{F}_{\mathrm{Laplace}}$	Pa	Laplace pressure	Equation (3.52)
$\boldsymbol{F}_{\mathrm{Marangoni}}$	Pa	Marangoni stress	Equation (3.53)
$\boldsymbol{f}$	m/s^2	body force per unit mass	
f	Hz	frequency	
GOR	–	Gained Output Ratio	Equation (2.3)
$\boldsymbol{g}$	m/s^2	gravitational acceleration	
g	–	as subscript indicates the gas	
G	–	as subscript indicates the gas	
H	$1/m$	mean curvature	Equation (3.47)
H	–	relative film thickness	Equation (4.32)
H_L	m	mean liquid height	Equation (3.106)
i	–	$\sqrt{-1}$	
i	–	integer index in a program	
i	–	as subscript indicates the gas-liquid interface	
I	–	as subscript indicates the gas-liquid interface	
$\boldsymbol{i}$	–	unit vector in x direction	
$\boldsymbol{I}$	–	identity matrix	
$\boldsymbol{j}$	–	unit vector in y direction	
$\boldsymbol{k}$	–	unit vector in z direction	
k	m^2/s^2	turbulent kinetic energy	Equation (3.72)
k_B	J/K	Boltzmann constant	
k_x	$1/m$	angular wavenumber	
K	m^{-2}	total curvature	Equation (3.48)
Ka	–	Kapitza number	Equation (4.2)
ℓ	m	size of a large turbulent eddy	Equation (3.81)
ℓ	m	mixing length	Equation (3.96)
ℓ^+	–	dimensionless mixing length	Equation (3.98)

Table F.2: (continued)

symbol	unit	explanation	see
l	–	as subscript indicates the liquid	
L	–	as subscript indicates the liquid	
L	m	length of a control volume	
L_{cut}	m	length from where the mesh was rectangular	Section 7.4
L	m	extension of a wire's wake	
L	m	distance between wires	
Le	–	Lewis number	Equation (3.41)
M	kg	mass	
$\dot{M}$	kg/s	mass flow rate	Equation (4.18)
m	kg	mass	
$\dot{m}$	$\text{kg}/(\text{m}^2\ \text{s})$	mass flux	
N	–	natural number	
N_{cells}	–	number of cells	
N_t	–	number of time steps	
N_x	–	number of cells in x direction	
N_y	–	number of cells in y direction	
Nu	–	Nusselt number	Equation (4.30)
$\boldsymbol{n}$	–	unit normal vector of the surface	
$\boldsymbol{P}$	kg m/s	momentum	
P	–	point in numerical grid	Figure 5.1
$\mathcal{P}$	m^2/s^3	production of turbulent kinetic energy	Equation (3.79)
Pr	–	Prandtl number	Equation (3.34)
Pr_t	–	turbulent Prandtl number	Section 3.4.5
p	Pa	pressure	
p_c	Pa	critical pressure	
p_v	Pa	partial pressure of the vapor	
Q	m^2/s^3	specific volumetric heat supply	
$\dot{Q}$	W	heat flow rate	Equation (4.20)
$\boldsymbol{q}$	W/m^2	heat flux	Equation (3.35)
q_w	W/m^2	wall heat flux	
q_u	m^2/s	specific flow rate per unit of the film width	Equation (4.68)
r	m	curvature radius	Equation (3.100)
r_{evap}	1/s/m	evaporation rate	Equation (9.73)
Re	–	Reynolds' number	Equation (4.1) (film)
Re_c	–	critical Reynolds' number	Equation (4.47)
Re_t	–	turbulent Reynolds' number	Equation (3.94)
R_u and R_p	depends	source terms in model equation	Equation (5.10)
s	–	as subscript indicates the gas-liquid interface, i. e., surface	
S	g/ℓ or kg/kg or g/kg	salinity	
S_k	m^2/s^3	sink of extension of k-ε model	Equation (3.110)
$\boldsymbol{S}_n$	Pa	normal surface stress	Equation (3.50)
$\boldsymbol{S}_s$	Pa	tangential surface stress	Equation (3.51)
Sc	–	Schmidt number	Equation (3.40)
T	K or °C	temperature	

Table F.2: (continued)

symbol	unit	explanation	see
T_c	K or °C	critical temperature	
T_I or T_i	K or °C	temperature of gas-liquid interface	
T_s	K or °C	temperature of surface (= T_i)	
T_{sat}	K or °C	saturation temperature	
T_w	K or °C	wall temperature	
t	s	time	
t_{period}	s	wave period	
t_{total}	s	total simulated time	Equation (3.83)
$\boldsymbol{U}$	m/s	velocity	
$\boldsymbol{U}_c$	m/s	compressing velocity	Section 5.4
U_r	m/s	residual velocity	Section 5.4
U^+	–	dimensionless velocity	Equation (3.86)
u	m/s	streamwise velocity	
u_I or u_i	m/s	u at gas-liquid interface	
u_s	m/s	u at surface (=interface)	
u_{s0}	m/s	u at surface of Nusselt film	Equation (4.35)
u_0	m/s	average velocity of Nusselt film	Equation (7.1)
V	m^3	volume	
v	m/s	cross-stream velocity	
v	–	as subscript indicates the vapor	
w	m/s	spanwise velocity	
w	–	western cell face	Figure 5.1
w	–	wall	
W	–	western neighbor of P	Figure 5.1
We	–	Weber number	Equation (4.45)
x	m	streamwise coordinate	
y	m	cross-stream coordinate	
y^+	–	dimensionless distance to the wall	Equation (3.87)
z	m	spanwise coordinate	
Z	m	width of the plate	

Table F.3: Table of symbols with Roman and Greek letters.

symbol	unit	explanation	see
$\varepsilon_{\mathrm{lw}}$	–	long wave parameter	Equation (4.76)
C_μ	–	model constant	Equation (3.76)
$\boldsymbol{F}_\Sigma$	N/m^3	Laplace volume force in Navier–Stokes equations	Equation (8.3)
f_μ	–	transition function	Equation (3.89)
ΔH	m	mean wave amplitude	Equation (3.105)
Δh	J/kg	latent heat of evaporation	
δ_{ij}	–	Kronecker delta	
δ_{in}	m	film thickness at inlet	
λ_x	m	wavelength	
Λ_{lw}	m	long wave length	Equation (4.76)
ν_t	m^2/s	turbulent viscosity	Equation (3.76)
δV	m^3	small volume	
κ^s	N s/m	surface viscosity	Section 3.2
μ^s	N s/m	surface viscosity	Section 3.2
ρ_v	kg/m^3	partial density of vapor	
Σ_k	m^2/s^3	other terms of k-ε model	Equation (3.110)
S_ε	m^2/s^4	source terms of extension of k-ε model	Equation (3.111)
t_ξ	s	time scale of largest turbulent eddies	Equation (3.108)
t_η	s	time scale of smallest turbulent eddies	Equation (3.109)
τ_w	Pa	wall shear stress	Equation (4.81)

Bibliography

[1] www.cfd-peric.de as accessed on 5th October 2018.

[2] www.openfoam.com as accessed on 22nd January 2019.

[3] https://www.advisian.com/en-gb/global-perspectives/the-cost-of-desalination as accessed on 12th September 2018.

[4] https://www.energyandcapital.com/report/water-desalination-investments/426 as accessed on 12th September 2018.

[5] Adomeit P and Renz U. Hydrodynamics of three-dimensional waves in laminar falling films. *International Journal of Multiphase Flow*, 26:1183–1208, 2000.

[6] Åkesjö A, Gourdon M, Vamling L, Innings F, and Sasic S. Hydrodynamics of vertical falling films in a large-scale pilot unit – a combined experimental and numerical study. *International Journal of Multiphase Flow*, 95:188–198, 2017.

[7] Al-Sibai F. *Experimentelle Untersuchung der Strömungscharakteristik und des Wärmeübergangs bei welligen Rieselfilmen*. PhD thesis, RWTH Aachen, 2004.

[8] Al-Sibai F, Leefken A, Lel V, and Renz U. Measurements of transport phenomena in thin wavy films. In *Second International Berlin Workshop on Transport Phenomena with Moving Boundaries (IBW2)*, 2003.

[9] Al-Sibai F, Leefken A, and Renz U. Local and instantaneous distribution of heat transfer rates through wavy films. In *First International Berlin Workshop on Transport Phenomena with Moving Boundaries (IBW1)*, 2001.

[10] Albert C. *On Stability of Falling Films: Numerical and Analytical Investigations*. PhD thesis, Technical University of Darmstadt, 2013. http://tuprints.ulb.tu-darmstadt.de/3665/ as accessed on 8th January 2019.

[11] Albert C, Raach H, and Bothe D. Influence of surface tension models on the hydrodynamics of wavy laminar falling films in volume of fluid-simulations. *International Journal of Multiphase Flow*, 43:66–71, 2012.

[12] Alekseenko S, Antipin V, Bobylev A, and Markovich D. Application of PIV to velocity measurements in a liquid film flowing down an inclined cylinder. *Experiments in Fluids*, 43:197–207, 2007.

[13] Alekseenko SV, Nakoryakov VE, and Pokusaev BG. *Wave Flow of Liquid Films*. Begell House, 1994.

[14] Alhusseini A, Tuzla K, and Chen JC. Falling film evaporation of single component liquids. *International Journal of Heat and Mass Transfer*, 41(12):1623–1632, 1998.

[15] Alke A. *Numerische Modellierung des Einflusses löslicher Tenside auf fluide Zweiphasensysteme*. Dissertation Universität Paderborn, Shaker, 2008.

[16] American Water Works Association. *Desalination of Seawater and Brackish Water*. AWWA, 2006.

[17] Anderson Jr JD. *Computational Fluid Dynamics (The Basics with Applications)*. McGraw-Hill, 1995.

[18] Aris R. *Vectors, Tensors, and the Basic Equations of Fluid Mechanics*. Dover, 1989.

[19] Ashgriz N and Poo JY. Flair: Flux line-segment model for advection and interface reconstruction. *Journal of Computational Physics*, 93:449–468, 1991.

[20] Badam VK. *Experimental and Theoretical Investigations of the Evaporation of Fluids from Free Surfaces*. PhD thesis, Universität Erlangen-Nürnberg, 2007.

[21] Bammel K, editor. *Faszination Physik*. Spektrum Akademischer Verlag, 2004.

[22] Bennett A. *Lagrangian Fluid Dynamics*. Cambridge University Press, 2006.

https://doi.org/10.1515/9783110592337-018

[23] Bernard PS and Wallace JM. *Turbulent Flow (Analysis, Measurement, and Prediction)*. John Wiley & Sons, 2002.

[24] Bird RB, Stewart WE, and Lightfoot EN. *Transport Phenomena*. John Wiley & Sons, second edition, 2002.

[25] Brackbill JU, Kothe DB, and Zemach C. A continuum method for modeling surface tension. *Journal of Computational Physics*, 100:335–354, 1992.

[26] Brandt S. *Datenanalyse*. Spektrum Akademischer Verlag, fourth edition, 1999.

[27] Brauer H. *Strömung und Wärmeübergang bei Rieselfilmen*. VDI-Forsch.-Heft 457, 1956.

[28] Bullard JW, Garboczi EJ, Carter WC, and Fuller Jr ER. Numerical methods for computing interfacial mean curvature. *Computational Material Science*, 4:103–116, 1995.

[29] CD adapco Group. *Methodology Star-CD Version 3.20*, 2004.

[30] Chang H-C and Demekhin EA. *Complex Wave Dynamics on Thin Films*. Elsevier, 2002.

[31] Chorin AJ. Flame advection and propagation algorithms. *Journal of Computational Physics*, 35:1–11, 1980.

[32] Christians M. *Falling Film Evaporation (Heat Transfer and Visualization)*. VDM Verlag Dr. Müller, 2011.

[33] Chun KR and Seban RA. Heat transfer to evaporating liquid films. *ASME Journal of Heat Transfer*, 93:391–396, 1971.

[34] Chung TJ. *Computational Fluid Dynamics*. Cambridge University Press, 2002.

[35] Davidson PA. *Turbulence (An Introduction for Scientists and Engineers)*. Oxford University Press, 2004.

[36] Demekhin EA and Shkadov VYa. On three-dimensional non-stationary waves in a falling liquid film (in Russian). *Akademiia Nauk SSSR, Izvestiia, Mekhanika Zhidkosti i Gaza*, 5:21–27, 1984.

[37] Deshpande SS, Anumolu L, and Trujillo MF. Evaluating the performance of the two-phase flow solver interfoam. *Computational Science & Discovery*, 5:014016 (36 pp.), 2012.

[38] Dietze GF. *Flow Separation in Falling Liquid Films*. Dissertation RWTH Aachen, Sierke Verlag, 2010.

[39] Dietze G, Lel VV, and Kneer R. Experimental investigation and modeling of wavy film flow. In *Third International Berlin Workshop on Transport Phenomena with Moving Boundaries (IBW3)*, 2005.

[40] Dietze GF, Rohlfs W, Nährich K, Kneer R, and Scheid B. Three-dimensional flow structures in laminar falling liquid films. *Journal of Fluid Mechanics*, 743:75–123, 2014.

[41] Doro EO. *Computational Modeling of Falling Liquid Film Free Surface Evaporation*. PhD thesis, Georgia Institute of Technology, 2012. https://smartech.gatech.edu/handle/1853/44812 as accessed on 14th February 2019.

[42] Durbin PA and Pettersson Reif BA. *Statistical Theory and Modeling for Turbulent Flows*. John Wiley & Sons, 2001.

[43] Edwards DA, Brenner H, and Wasan DT. *Interfacial Transport Processes and Rheology*. Butterworth-Heinemann, 1991.

[44] El-Dessouky HT and Ettouney HM. *Fundamentals of Salt Water Desalination*. Elsevier, 2002.

[45] Elimelech M and Phillip WA. The future of seawater desalination: Energy, technology, and the environment. *Science*, 333, August 2011.

[46] Emad V. *Evaluating the Performance of Various Convection Schemes on Free Surface Flows by Using Interfoam Solver*. Master's thesis, Eastern Mediterranean University, 2014.

[47] Evans G, Blackledge J, and Yardley P. *Numerical Methods for Partial Differential Equations*. Springer Undergraduate Mathematics Series, 2000.

[48] Faghri A and Zhang Y. *Transport Phenomena in Multiphase Systems*. Elsevier Academic Press, 2006.

[49] Ferreira VG, Mangiavacchi N, Tomé MF, Castelo A, Cuminato JA, and McKee S. Numerical simulation of turbulent free surface flow with two-equation k-ϵ eddy-viscosity models. *International Journal for Numerical Methods in Fluids*, 44:347–375, 2004.

[50] Ferziger JH and Peric M. *Computational Methods for Fluid Dynamics*. Springer, third edition, 2002.

[51] Fröhlich J. *Large Eddy Simulation turbulenter Strömungen*. Teubner, 2006.

[52] Gad el Hak M and Tsai HM, editors. *Transition and Turbulence Control*. World Scientific, 2006.

[53] Gambaryan-Roisman T and Stephan P. Stability of falling films on grooved surfaces. In *First International Berlin Workshop on Transport Phenomena with Moving Boundaries (IBW1)*, 2001.

[54] Gambaryan-Roisman T and Stephan P. Gravity- and shear-driven liquid films on grooved surfaces: heat transfer and flow stability analysis. In *Second International Berlin Workshop on Transport Phenomena with Moving Boundaries (IBW2)*, 2003.

[55] Gambaryan-Roisman T and Stephan P. Film flow down vertical and inclined microstructured heated walls: hydrodynamics, heat transfer and stability. In *Third International Berlin Workshop on Transport Phenomena with Moving Boundaries (IBW3)*, 2005.

[56] Gatignol R and Prud'homme R. *Mechanical and Thermodynamical Modeling of Fluid Interfaces*. World Scientific, 2001.

[57] Gatski TB, Hussaini MY, and Lumley JL, editors. *Simulation and Modeling of Turbulent Flows*. Oxford University Press, 1996.

[58] Gropp W, Lusk E, and Skjellum A. *Using MPI*. MIT Press, 1995.

[59] Gusmano G, Montesperelli G, Forte G, and Olzi E. On-line corrosion resistance tests in sea water on metals for MED plants. *Desalination*, 183:187–194, 2005.

[60] Hackbusch W. *Iterative Solution of Large Sparse Systems of Equations*. Springer, 1993.

[61] Haggstrom P. The Laplacian in curvilinear coordinates – the full story. https://gotohaggstrom.com/calculus.html as accessed on 18th March 2019.

[62] Hagiwara T and Madarame H. Two-dimensional numerical analysis of turbulent flow with a free surface in a tank. *JSME International Journal/Series II*, 35(2):174–178, 1992.

[63] Harvie DJE, Davidson MR, and Rudman M. An analysis of parasitic current generation in volume of fluid simulations. *Australian and New Zealand Industrial and Applied Mathematics Journal*, 46(E):C133–C149, 2005.

[64] Hewitt GF and Vassilicos JC, editors. *Prediction of Turbulent Flows*. Cambridge University Press, 2005.

[65] Hirt CW and Nichols BD. Volume of fluid (VOF) method for the dynamics of free boundaries. *Journal of Computational Physics*, 39:201–225, 1981.

[66] Hopp V. *Wasser (Krise?)*. Wiley-VCH, 2004.

[67] International Desalination Association, September 2018. idadesal.org/desalination-101/desalination-by-numbers.

[68] International Water Association, September 2018. www.iwa-network.org/desalination-past-present-future.

[69] Juric D and Tryggvason G. Computation of boiling flows. *International Journal of Multiphase Flow*, 24:387–410, 1998.

[70] Kafi F, Renaudin V, Alonso D, and Hornut JM. New MED plate desalination process: Thermal performances. *Desalination*, 166:53–62, 2004.

[71] Kafi F, Renaudin V, Alonso D, Hornut JM, and Weber M. Experimental study of a three-effect plate evaporator: seawater tests in La Spezia. *Desalination*, 182:175–186, 2005.

[72] Kalliadasis S, Ruyer-Quil C, Scheid B, and Velarde MG. *Falling Liquid Films*. Springer, 2012.

[73] Kanzow C. *Numerik linearer Gleichungssysteme: Direkte und iterative Verfahren*. Springer, 2004.

[74] Kapitza PL. Wave flow of thin layer of viscous fluid (in Russian). *Journal Experimental Theoretical Physics (UdSSR)*, 18:3–18, 1948.

[75] Kapitza PL. Wave flow of thin layer of viscous fluid (in Russian). *Journal Experimental Theoretical Physics (UdSSR)*, 19:105–120, 1949.

[76] Kharangate CR. *Experimental, Theoretical and Computational Modeling of Flow Boiling, Flow Condensation and Evaporating Falling Films*. PhD thesis, Purdue University, 2016. https://docs.lib.purdue.edu/open_access_dissertations/783 as accessed on 20th March 2019.

[77] Kharangate CR, Lee H, and Mudawar I. Computational modeling of turbulent evaporating falling films. *International Journal of Heat and Mass Transfer*, 81:52–62, 2015.

[78] Kolev NI. *Multiphase Flow Dynamics 1 (Fundamentals)*. Springer, 2002.

[79] Kolev NI. *Multiphase Flow Dynamics 2 (Mechanical and Thermal Interactions)*. Springer, 2002.

[80] Krahl R, Adamov M, Lozano Avilés M, and Bänsch E. A model for two phase flow with evaporation. *WIAS preprint 899*, 2004.

[81] Kunugi T and Kino C. DNS of falling film structure and heat transfer via MARS method. *Computers and Structures*, 83:455–462, 2005.

[82] Köckler N and Schwarz HR. *Numerische Mathematik*. Teubner, fifth edition, 2004.

[83] Lafaurie B, Nardone C, Scardovelli R, Zaleski S, and Zanetti G. Modelling merging and fragmentation in multiphase flows with SURFER. *Journal of Computational Physics*, 113:134–147, 1994.

[84] Launder BE and Spalding DB. The numerical computation of turbulent flows. *Computer Methods in Applied Mechanics and Engineering*, 3:269–289, 1974.

[85] Leefken A and Renz U. Numerical analysis on hydrodynamics and heat transfer of laminar-wavy falling films. In *First International Berlin Workshop on Transport Phenomena with Moving Boundaries (IBW1)*, 2001.

[86] Lemos CM. A simple numerical technique for turbulent flows with free surfaces. *International Journal for Numerical Methods in Fluids*, 15:127–146, 1992.

[87] Lesieur M, Métais O, and Comte P. *Large-Eddy Simulations of Turbulence*. Cambridge University Press, 2005.

[88] Lien FS, Chen WL, and Leschziner MA. Low-Reynolds-number eddy-viscosity modelling based on non-linear stress-strain/vorticity relations. In *Proc. 3rd Symp. on Engineering Turbulence Modelling and Measurements*, Crete, Greece, 1996.

[89] Limberg H. Über die turbulente Strömung in einem Rieselfilm. *Monatsberichte der DAW Berlin*, 12(5):333–341, 1970.

[90] Loeb S and Sourirajan S. Seawater demineralization by means of an osmotic membrane. In *Water Conversion II Advances in Chemistry, Series Number 28*. American Chemical Society, Washington, DC, 1963.

[91] Lorch W, editor. *Handbook of Water Purification*. John Wiley & Sons, 1987.

[92] Lorencez C, Nasr-Esfahany M, Kawaji M, and Ojha M. Liquid turbulence structure at a sheared and wavy gas–liquid interface. *International Journal of Multiphase Flow*, 23(2):205–226, 1997.

[93] Lozano Aviles M. *Experiments on Falling Film Evaporation of a Water-Ethylene Glycol Mixture on a Surface with Longitudinal Grooves*. PhD thesis, Technical University Berlin, 2007. https://d-nb.info/984542469/34 as accessed on 11th February 2019.

[94] Lötstedt P. A front tracking method applied to Burger's equation and two-phase porous flow. *Journal of Computational Physics*, 47:211–228, 1982.

[95] MacHarg JP. Retro-fitting existing SWRO systems with a new energy recovery device. *Desalination*, 153:253–264, 2002.

[96] Merker GP and Eiglmeier C. *Fluid- und Wärmetransport: Wärmeübertragung*. Teubner, 1999.

[97] Mills AF and Chung DK. Heat transfer across turbulent falling films. *International Journal of Heat and Mass Transfer*, 16:694–696, 1973.

[98] Mitrovic J. Filmkondensation ruhender Dämpfe an senkrechten Flächen. *BWK*, 42:598–609, 1990.

[99] Miyara A. Numerical analysis on flow dynamics and heat transfer of falling liquid films with interfacial waves. *Heat and Mass Transfer*, 35:298–306, 1999.

[100] Miyara A. Numerical simulation of wavy liquid film flowing down on a vertical wall and an inclined wall. *International Journal of Thermal Sciences*, 39:1015–1027, 2000.

[101] Miyara A. Numerical simulation of heat and mass transfer of wavy falling films. In *Second International Berlin Workshop on Transport Phenomena with Moving Boundaries (IBW2)*, 2003.

[102] Mudawwar IA and El-Masri MA. Momentum and heat transfer across freely-falling turbulent liquid films. *International Journal of Multiphase Flow*, 12:771–790, 1986.

[103] Munson BR, Young DF, and Okiishi TH. *Fundamentals of Fluid Mechanics*. John Wiley & Sons, fourth edition, 2002.

[104] Noh WF and Woodward P. SLIC (Simple Line Interface Calculations). In *Fifth International Conference on Numerical Methods in Fluid Dynamics*, Enschede, The Netherlands, June 1976. https://www.osti.gov/servlets/purl/7261651-xagqbi/ as accessed on 6th October 2018.

[105] Nosoko T, Yoshimura PN, Nagata T, and Oyakawa K. Characteristics of two-dimensional waves on a falling liquid film. *Chemical Engineering Science*, 51:725–732, 1996.

[106] Numrich R. Heat transfer in turbulent falling films. *Chemical Engineering & Technology*, 18:171–177, 1995.

[107] Nußelt W. Die Oberflächenkondensation des Wasserdampfes. *Zeitschrift VDI*, 60(27):541–546, 1916.

[108] Pacheco PS. *Parallel Programming with MPI*. Kaufmann, 1997.

[109] Papon P, Leblond J, and Meijer PHE. *The Physics of Phase Transitions*. Springer, 2002.

[110] Pope SB. *Turbulent Flows*. Cambridge University Press, 2000.

[111] Portalski S. Eddy formation in film flow down a vertical plate. *Industrial & Engineering Chemistry Fundamentals*, 3(1):49–53, 1964.

[112] Portalski S. Velocities in film flow of liquids on vertical plates. *Chemical Engineering Science*, 19:575–582, 1964.

[113] Press WH, Teukolsky SA, and Vetterling WT. *Numerical recipes in FORTRAN77*. Cambridge University Press, second edition, 1992.

[114] Pressley A. *Elementary Differential Geometry*. Springer, 2001.

[115] Raach H. Extension of k-ϵ models for a turbulent falling film. *Heat and Mass Transfer*, 48:1231–1234, 2012.

[116] Raach H and Mitrovic J. Seawater falling film evaporation on vertical plates with turbulence wires. *Desalination*, 183:307–316, 2005.

[117] Raach H and Mitrovic J. Simulation of heat and mass transfer in a multi-effect distillation plant for seawater desalination. *Desalination*, 204:416–422, 2007.

[118] Raach H and Somasundaram S. Numerical investigations on heat transfer in falling films around turbulence wires. In *Proceedings of the International Conference "Eurotherm 2008"*, Eindhoven, Netherlands, 2008.

[119] Raach H, Somasundaram S, and Mitrovic J. Optimisation of turbulence wire spacing in falling films performed with OpenFOAM. *Desalination*, 267:118–119, 2011.

[120] Reif F. *Fundamentals of Statistical and Thermal Physics*. McGraw-Hill, 1965.

[121] Renaudin V, Kafi K, Alonso D, and Andreoli A. Performances of a three-effect plate desalination process. *Desalination*, 182:165–173, 2005.

[122] Rider WJ and Kothe DB. Reconstructing volume tracking. *Journal of Computational Physics*, 141:112–152, 1998.

[123] Rieber M. *Numerische Modellierung der Dynamik freier Grenzflächen in Zweiphasenströmungen*. PhD thesis, University of Stuttgart, 2004.

[124] Rudman M. Volume-tracking methods for interfacial flow calculations. *International Journal for Numerical Methods in Fluids*, 24:671–691, 1997.

[125] Rusche H. *Computational Fluid Dynamics of Dispersed Two-Phase Flows at High Phase Fractions*. PhD thesis, Imperial College London/University of London, 2002. https://spiral.imperial.ac.uk:8443/handle/10044/1/8110 as accessed on 5th October 2018.

[126] Scheid B, Ruyer-Quil C, and Manneville P. Wave patterns in film flows: modelling and three-dimensional waves. *Journal of Fluid Mechanics*, 562:183–222, 2006.

[127] Service RF. Desalination freshens up. *Science*, 313, August 2006.

[128] Sharqawy MH, Lienhard V JH, and Zubair SM. The thermophysical properties of seawater: A review of existing correlations and data. *Desalination and Water Treatment*, 16:354–380, 2010.

[129] Shikhmurzaev YD. *Capillary Flows with Forming Interfaces*. Chapman & Hall/CRC, 2008.

[130] Slattery JC, Sagis L, and Oh E-S. *Interfacial Transport Phenomena*. Springer, second edition, 2007.

[131] Somasundaram S. *Validierung von OpenFOAM für wellige, laminare Fallfilme*. große Studienarbeit, University of Paderborn, 2007.

[132] Somasundaram S. *Numerische Untersuchung des Wärmeübergangs in Fallfilmen mit Stolperdrähten*. Master's thesis, University of Paderborn, 2007.

[133] Spiegler KS and Laird ADK, editors. *Principles of Desalination (Part A and B)*. Academic Press, second edition, 1980.

[134] Stone HL. Iterative solution of implicit approximations of multidimensional partial differential equations. *SIAM Journal of Numerical Analysis*, 5(3):530–558, 1968.

[135] Stover RL. Seawater reverse osmosis with isobaric energy recovery devices. *Desalination*, 203:168–175, 2007.

[136] Tihon J. Hydrodynamics of the solitary waves travelling down a liquid film. In *Second International Berlin Workshop on Transport Phenomena with Moving Boundaries (IBW2)*, 2003.

[137] Tsinober A. *An Informal Introduction to Turbulence*. Kluwer Academic Publishers, 2001.

[138] Tsinober A. *An Informal Conceptual Introduction to Turbulence*. Springer, 2009.

[139] Ubbink O. *Numerical Prediction of Two Fluid Systems with Sharp Interfaces*. PhD thesis, Imperial College London/University of London, 1997. http://powerlab.fsb.hr/ped/kturbo/OpenFOAM/docs/OnnoUbbinkPhD.pdf as accessed on 5th October 2018.

[140] United States Army. *Water Desalination*. University Press of the Pacific, 2005 (official credit: Department of the Army, USA, Technical Manual TM 5-813-8, Water Desalination).

[141] Unterberg W and Edwards DK. Evaporation from falling saline water films in laminar transitional flow. *AIChE Journal*, 11(6):1073–1081, 1965.

[142] VDI. *VDI-Wärmeatlas*. Springer, tenth edition, 2006.

[143] Versteeg HK and Malalasekera W. *An Introduction to Computational Fluid Dynamics (The Finite Volume Method)*. Pearson-Prentice Hall, 1995.

[144] Versteeg HK and Malalasekera W. *An Introduction to Computational Fluid Dynamics (The Finite Volume Method)*. Pearson-Prentice Hall, second edition, 2007.

[145] Voutchkov N. *Desalination Engineering: Planning and Design*. McGraw-Hill, 2013.

[146] Voutchkov N. *Desalination Engineering: Operation and Maintenance*. McGraw-Hill, 2014.

[147] Wadekar V. Modelling and prediction of falling film evaporation. In *Second International Berlin Workshop on Transport Phenomena with Moving Boundaries (IBW2)*, 2003.

[148] Water Reuse Association. Water desalination costs, January 2012. White Paper.

[149] Welch SWJ and Wilson J. A Volume of Fluid based method for fluid flow with phase change. *Journal of Computational Physics*, 160:662–682, 2000.

[150] Weller HG. *A New Approach to VOF-Based Interface Capturing Methods for Incompressible and Compressible Flow*. Technical report, OpenCFD, 2008.

[151] Woldai A. *Multi-Stage Flash Desalination: Modeling, Simulation, and Adaptive Control*. CRC Press, 2017.

[152] Youngs DL. Time-dependent multi-material flow with large fluid distortion. In Morton KW and Baines MJ, editors, *Numerical Methods for Fluid Dynamics*, pages 273–285. Academic, New York, 1982.

[153] Zheng H. *Solar Energy Desalination Technology*. Elsevier, 2017.

[154] Zienkiewicz OC and Morgan K. *Finite Elements and Approximation*. Dover, 2006.

[155] Zienkiewicz OC, Taylor RL, and Nithiarasu P. *The Finite Element Method for Fluid Dynamics*. Butterworth Heinemann, sixth edition, 2005.

[156] Zienkiewicz OC, Taylor RL, and Zhu JZ. *The Finite Element Method: Its Basis and Fundamentals*. Butterworth Heinemann, sixth edition, 2005.

Index

Bei Fragen zur Produktsicherheit wenden Sie sich bitte an:
If you have any questions regarding product safety,
please contact:

Walter de Gruyter GmbH
Genthiner Straße 13
10785 Berlin
productsafety@degruyterbrill.com